RISK AND CRISIS COMMUNICATIONS

Methods and Messages

by Pamela (Ferrante) Walaski

A JOHN WILEY & SONS, INC., PUBLICATION

Published by John Wiley & Sons, Inc., Hoboken, New Jersey
Published simultaneously in Canada

For general information on our other products and services or for technical support, please contact our Customer Care Department within the United States at (800) 762-2974, outside the United States at (317) 572-3993 or fax (317) 572-4002.

Wiley also publishes its books in a variety of electronic formats. Some content that appears in print may not be available in electronic formats. For more information about Wiley products, visit our website at www.wiley.com.

Library of Congress Cataloging-in-Publication Data

Walaski, Pamela, 1959-
  Risk and crisis communications : methods and messages / Pamela Walaski.
    p. cm.
  Includes index.
  ISBN 978-0-470-59273-1 (cloth)
  1. Risk communication.   2. Health risk communication.   3. Crisis management.   4. Public safety.   5. Industrial safety.   I. Title.
  T10.68.W35 2011
  363.3401'4–dc22

                                                                                          2011008252

Printed in Singapore

oBook ISBN: 978-1-118-09342-9
ePDF ISBN: 978-1-118-09345-0
ePub ISBN: 978-1-118-09344-3

10 9 8 7 6 5 4 3 2 1

CONTENTS

LIST OF TABLES

PREFACE

I wish I could take credit for the idea to write this book, but I can't. Several years ago, I was in my office putting the finishing touches on some slides for an upcoming conference session on risk and crisis communications at the annual Professional Development Conference of the American Society of Safety Engineers (ASSE) in San Antonio, Texas. An e-mail popped up in my inbox from an unknown person, who happened to be Bob Esposito, an associate publisher at John Wiley & Sons. He noted that he had seen and heard about several other sessions where I had presented on this topic at other major national conferences and wondered if I had ever thought about writing a book about it. While the idea of writing a book was one of my long-term goals, it never occurred to me that a publisher would think me ready to write one now. I had just finished writing a chapter for a book published by ASSE and had also continued the practice of publishing articles for print and online newsletters for several different organizations and associations. While I enjoyed writing and had been pleased with the articles that I had published thus far, writing an entire book was not something I thought I was ready for.

And to this day, I'm still not sure I was ready for it. As I have joked several times over the past 24 months, writing a 10,000-word chapter or a 2,500-word newsletter article is actually pretty easy and had become easier each time I penned one. However, a book with nearly 100,000 words turned out to be an incredibly daunting task—one that has, at turns, energized me, challenged me, and beaten me down, sometimes during the same writing session. When Bob asked me to write this book, I was flattered and, without really understanding what it would take to get it done, said yes. While I never doubted that I had much to say on this topic and believed that my ideas and those of others that I have collected and utilized for this book would be beneficial to my fellow safety, health, and environmental (SH&E) professionals, there have been times when I have had the (probably) universal tinge of doubt about whether or not what I had to say would be interesting to anyone but me.

This book is not intended to be a significant seminal work on the topic of risk and crisis communications. It is, and always was, intended to bring the topic down to the level of general safety practitioners who are looking to add more value to their professional skill set and ultimately to their employers. It is written for safety professionals who, like me, are passionate about what they do and want to better understand how to bring the message of safety to the audiences that make up their specific work environment. This book is filled with general concepts, theories, and practical applications that can be used by anyone in the field with some basic management responsibilities, and even by those with no management level tasks who simply want to broaden their knowledge of this particular aspect of the practice of safety.

Everyone always has a list of people to thank and I am no exception. My first thanks go to Bob Esposito who, I think, took a bit of a leap with me and gave me a shot. My second thanks go to those who have been working in the area of risk and crisis communications for many more years than I and who have contributed greatly to the understanding of the practice. They include Peter Sandman, Vincent Covello, Kathleen Fearn-Banks, Regina Lundgren, and Andrea McMakin, whose body of work has contributed greatly to the pages that follow. I also want to thank Morgan Kelly for her invaluable assistance with designing some of the illustrations.

I also need to mention my children, Jason and Chloe, who have allowed me to hone my parenting skills on them. I am cautiously optimistic that I have gotten better at it over time; and I have learned more from them than I think they understand at this stage of their lives, although when they are parents, I do believe they will get it as well. And finally, yet most importantly, to my husband Jeff, the guy who fell in love with me while I was knee-deep in writing this book. Over the year and a half of our courtship and the early months of our marriage, he never failed to take a back seat without complaining when I needed to spend just one more evening (I promise!) or one more Saturday (I swear!) writing and editing it. When I almost gave up, he encouraged me to keep at it; and during the long hours of research, writing, and editing, he was always an unwavering guardian of my time, making sure the task never overwhelmed me. His patience, devotion, love, and support are gifts that I hold close in my heart.

PAMELA (FERRANTE) WALASKI

1

INTRODUCTION

More than 30 years ago, a seminal event in the field of crisis communications occurred at a nuclear power plant operated by Metropolitan Edison in Middletown, Pennsylvania, just outside of the state capital of Harrisburg. The plant, known as Three Mile Island (TMI), was the scene of an incident involving a stuck valve that resulted in the partial meltdown of a nuclear reactor. While TMI was not a serious accident in terms of human fatalities or injuries or release of dangerous radioactivity, it did identify serious gaps in the nuclear industry's ability to communicate during critical events and led to the establishment of the Kemeny Commission, whose tasks included writing recommendations on how nuclear utilities should improve their ability to communicate in the event of an accident.

As a young undergraduate student attending Shippensburg State College (now University), just outside of Harrisburg and Middletown, I remember the difficulties we had in understanding what was happening and how it might affect us at that time and in the future. Living in a college dormitory equipped with pay phones only in the main lobby and one television set for the entire residence of 200-plus students, the methods of communication available to let us know what was happening were extremely limited, leaving us in the dark, while National Guard troops pulled up on our campus as we prepared to take in evacuees.

Risk and Crisis Communications: Methods and Messages, First Edition. Pamela (Ferrante) Walaski.
© 2011 John Wiley & Sons, Inc. Published 2011 by John Wiley & Sons, Inc.

Imagine how that lack of information would play out if TMI happened today. Our current culture and society relies increasingly on written and verbal messages on a near-constant basis to evaluate the world and the risks associated with living in it. These messages do more than simply provide information; they can cause large groups of people to behave in certain ways as well as change their perceptions of the world around them. As part of their functional responsibilities, safety, health, and environmental (SH&E) professionals are being called upon more frequently to develop the means and the messages to assess and communicate risks to the audiences of their organizations that include their internal workforce, the general public, vendors, suppliers, and other organizations within their field.

Risk and crisis communications is a process of communicating information by a public or private organization to an audience. The information is typically communicated following a formal or informal risk assessment process that delineates hazards that may occur to the organization and require some level of knowledge imparted to the audience on how the hazards will impact them and how they can prepare for the hazard. The process most often occurs when hazards are already occurring, are about to occur, or being planned for as part of an overall emergency response preparedness process.

In most literature, the terms "risk communications" and "crisis communications" are used to describe both the process of developing a relationship with key audiences in which information is communicated about the hazard, as well as the specific messages that are crafted and delivered by various organizational representatives. Risk communications is most often the process and the messages that occur prior to the occurrence of a hazard. Risk communications helps audiences understand their risk as well as what activities they can undertake to prepare for the hazard situation. Crisis communications is the process and messages that are delivered at times of high stress, either because the hazard is already occurring or is imminent.

This book provides readers with a fundamental understanding of the process of developing and delivering risk and crisis communications and has been written to provide a means for SH&E professionals to develop a foundational understanding of risk and crisis communications and use that information to assess the needs of their organization.

In recent years the roles of SH&E professionals have been expanding into new and different arenas. SH&E professionals need to provide value to their organizations by increasing their skill set and the roles they can play in the overall functioning of the organization. The ability to do so will provide a key to their success, both individually and to the profession as a whole. This need to become more valuable to an organization is coupled with the increasing role of media and communication methods in the provision of information to the public at large. Organizations must respond to this need for information in a way that is accurate and timely and is structured in such a way as to be successful. This book will provide the information SH&E professionals need to assure their success in this process.

While the bulk of the earliest history of the development of formal risk and crisis communications techniques centered on the environmental remediation and clean-up arena, more recent efforts have broadened into multiple arenas. The use of the techniques and activities described in this book, as well as others, now commonly cover events such as natural disasters; security incidents; public health crises; and workplace catastrophes, including fatalities and major incidents. Some threads of the theoretical foundations of risk and crisis communications can even be woven into much of occupational safety and health training classes that occur in just about every workplace.

The methods used to communicate risk and crisis information also vary from oral methods such as press conferences, broadcast interviews, public meetings, and safety meetings. Written communication methods range from the traditional press release to brochures, safety posters, and newsletters. Newer methods include robocalls, podcasts, websites, blogs, and social networking sites.

This book will take the reader through the fundamentals of risk and crisis communications and begins by providing a common set of working definitions for a variety of terms used throughout the book, including "risk," "crisis," "risk communications," and "crisis communications." Later chapters review the current theoretical foundations that have been developed by such leading experts in the fields such as Vincent Covello and his colleagues at The Center for Risk Communication; Peter Sandman; and Regina Lundgren and Andrea McMakin. Some limited review of research conducted to ascertain the validity of risk and crisis messages will also be addressed.

Information will also be presented that will guide readers through the steps of developing risk and crisis messages, including understanding the constraints of the organization; the audience and communication topic; the goals and objectives of the messages; how to profile the intended audience; and how to successfully deal with strong audience emotions such as anger, mistrust, fear, and apathy. Additional information on avoiding common mistakes made during risk and crisis communication situations will be identified.

Several chapters will address the crafting of the actual messages that are delivered and will include two current techniques for message crafting: influence diagrams, developed by M. Granger Morgan at Carnegie Mellon University in Pittsburgh, Pennsylvania; and message maps, developed by Vincent Covello at The Center for Risk Communication. Vital for the successful delivery of messages is the ability to demonstrate empathy to the audience. This will be discussed as will the method for choosing an effective organizational spokesperson.

Other chapters will look at several models for crisis communications plans that can be used as templates for an organization, which will allow it to be prepared to respond quickly to crisis situations. Working with the media, including successful press conferences and media interviews, will be covered as well. Finally, the use of crisis communications when an organization encounters a fatality or a rumor will be examined in greater detail.

The text will close with case studies of two recent public events that provide a wealth of information to study the actual process: the worldwide H1N1 pandemic of 2009–2010 and the BP Deepwater Horizon explosion and oil spill of 2010, which killed 11 oil rig workers and caused unprecedented environmental damage throughout the Gulf of Mexico. The case studies will analyze comments made by various organizational representatives, politicians, and governmental authorities with regard to some of the theoretical foundations and message-crafting concepts discussed in the earlier portions of the text. Comments will be given related to the relative success or failure of the messages delivered. A final chapter will summarize the entire text and offer closing comments.

It is hoped that this text will provide readers with a solid foundation and increase their skill set for immediate use within their organizations. It will also provide additional questions that may lead to a more in-depth study of the topic by reading some of the reference material used.

2

GENERAL CONCEPTS OF RISK AND CRISIS COMMUNICATIONS

This chapter will provide a general understanding of the field of risk and crisis communications by looking at the history of the practice. Subsequently, some working definitions will be provided to establish common terms that will be used throughout the remainder of the book. Additional sections will review a basic model of communication and the chapter will end with some comments about the purpose and objectives of risk and crisis communication events.

HISTORICAL BACKGROUND

The use of oral and written communication techniques can be dated as far back as human existence and were used to communicate a variety of needs, wishes, commands, and information, including the types of risks humans were exposed to and as a means to warn of impending crises. However, the science, practice, and specific techniques that form what is currently known as risk and crisis communications have a much shorter history. Not much has been written to accurately date the earliest forms of risk and crisis communications, but many in the field would hesitate to go back any further than 25 or 30 years. Attributions often identify the introduction of the World Wide Web

Risk and Crisis Communications: Methods and Messages, First Edition. Pamela (Ferrante) Walaski.
© 2011 John Wiley & Sons, Inc. Published 2011 by John Wiley & Sons, Inc.

Fig. 2.1. William Ruckelshaus chairs the new Puget Sound Partnership leadership council, seeking to protect and restore Puget Sound (Steve Ringman/The Seattle Times)

and other forms of digital communication as a turning point in the need to provide messages to audiences that help them understand the risks of their lives due to the sheer speed of Internet messages as well as the substantial increase in the volume and type of messages available to the general public (Fearn-Banks 2007).

Much of what are considered contemporary risk and crisis communication activities have stemmed from environmental clean-up efforts that began in the United States with legislation in the 1980s. Vincent Covello and Richard Peters at The Center for Risk Communication in New York, along with David McCallum from Focus Group in Maryland, trace the terms "risk communications" and "crisis communications" and their widespread use back to William Ruckelshaus (Fig. 2.1), the first administrator of the U.S. Environmental Protection Agency (EPA), a federal agency formed in 1970 by President Richard Nixon (Peters *et al.* 1997).

Ruckelshaus's first term, which lasted until 1973, was noted for the development of the organizational structure of the EPA as well as its initial enforcement actions. But it was when Ruckelshaus returned to the EPA in 1983 that his mandates to inform and involve the public in decisions about environmental remediation and clean-up activities through the use of risk-based decision making set the stage for much of what is now common community involvement practice by the EPA and other governmental agencies (U.S. EPA 2009).

Subsequent major federal legislative efforts continued to require bureaucrats to involve the public in decision making. The 1986 Superfund Amendments and Reauthorization Act (SARA) added new requirements for such activity under the Superfund program and amended certain existing provisions to reinforce them. SARA also created a major public-private planning process for responding to emergency hazardous materials incidents under the Emergency Planning and Community Right to Know Act of 1986, also known as Title III. The planning process is organized and managed by state emergency response committees (SERCs), whose members are appointed by the governor and who oversee the state's response to public and private hazardous materials emergencies, and by local emergency planning committees (LERCs), whose members are a broad-based representation from governmental agencies, local response organizations and community groups.

KEY DEFINITIONS

In order to proceed with a logical discussion of risk and crisis communications, definitions for terms commonly used throughout this book are essential. This foundation will assist the reader in understanding the framework for all that follows. The two key terms in this book include "risk" and "crisis." Most definitions of "risk" incorporate the concept of a hazard that might or might not occur, along with an understanding of the severity of the hazard and the probability of its occurrence. Here are four definitions that combine these concepts:

From the American National Standard for Occupational Health and Safety Management Systems (ANSI/AIHA Z10-2005): "The identification and analysis, either qualitative or quantitative, of the likelihood of the occurrence of a hazardous event or exposure, and the severity of injury or illness that may be caused by it."

From the *Framework for Environmental Health Risk Management* (Presidential/ Congressional Commission on Risk Assessment and Risk Management, 1997): "Risk is defined as the probability that a substance or situation will produce harm under specified conditions. Risk is a combination of two factors: (1) the probability that an adverse event will occur and (2) the consequences of the adverse event."

And, finally, from *The Safety Professionals Handbook* (Fields 2008): "Risk is the probability (or likelihood) that a harmful consequence will occur as a result of an action."

Similar to the above definitions of risk, those for "crisis" share several common elements as well, including the idea that the threat posed by the event is at least as serious, and often catastrophic, and that the actual timing of the event is often unpredictable though not necessarily unexpected. Here are three definitions from three different books on the topic of risk and crisis communications:

- "A crisis can be defined as an event that is an unpredictable, major threat that that can have a negative effect on the organization, industry, or stakeholders if handled improperly. A crisis is unpredictable but not unexpected" (Coombs 1999).

- "A crisis is a major occurrence with a potentially negative outcome affecting the organization, company, or industry, as well as its publics, products, services, or good name. A crisis interrupts the normal business transactions and can sometimes threaten the existence of an organization" (Fearn-Banks 2007).

• "A crisis is a turning point that will decisively determine an outcome" (Lundgren and McMakin 2004).

Risk management and crisis management are terms that include common themes of evaluation and control of the risk or crisis in order to bring about a successful outcome or at least to minimize the damage from the event. Here Lundgren and McMakin say that risk management is "evaluating and deciding how to cope with a risk" (Lundgren and McMakin 2004), while Fearn-Banks (2007) notes that crisis management is "a process of strategic planning for a crisis...that removes some of the risk and uncertainty from the negative occurrence and thereby allows the organization to be in greater control of its own destiny." And finally, Coombs (1999) says crisis management is "a set of factors designed to combat crises and lessen the actual damage inflected by the crisis."

Communicating with stakeholders about risks is an interactive process that takes time and resources in order to be effective, as noted by Lundgren and McMakin. Their definition of risk communication focuses on the importance of interaction between the communicator and the audience, even if the goal of the interaction is only to solicit simplified information rather than considered opinions and suggested solutions for managing the risks. In addition, risk communication is an integral part of risk management as noted here by the U.S. Department of Health and Human Services (2006): "Risk communication: An interactive process of exchange of information and opinion among individuals, groups, and institutions; often involves multiple messages abut the nature of risk or expressing concerns, opinions, or reactions to risk messages or to legal and institutional arrangements for risk management."

Risk communication is also a process that is based upon scientific principles and theoretical foundations, particularly about audience perception of risk (see Chapter 3). Covello and his colleagues at The Center for Risk Communication have contributed significantly to the scientific principles upon which much of current risk and crisis communications are based. They define risk communication as "a science-based approach to communicating effectively in high concern situations" (Covello 2008).

Communicating during crises involves similar elements as are noted above. It is an interactive process between a communicator and an audience to transfer information about the crisis, which is an integral part in managing the crisis to lessen the severity of the outcome at the crisis's conclusion (Lundgren and McMakin 2004; Fearn-Banks 2007).

The foundations of crisis communications also have substantial theoretical foundations, provided by Covello and his colleagues, Fearn-Banks, and Peter Sandman (Sandman was the director of the Environmental Communication Research Program at Rutgers University from 1986 to 1992). During that time, the program published numerous scholarly articles and books on the subject of risk and crisis communications, which provided the foundational elements used by many current practitioners of risk and crisis communications. Following his work at Rutgers, Sandman left to begin a consulting practice and has become one of the preeminent experts in the field (Sandman 2009). Sandman postulates that crisis communications is a subset of risk communications, performed by communicators when the hazard facing the stakeholder is high and a

TABLE 2.1. Differences between Risk Communications and Crisis Communications

Risk Communications	Crisis Communications
• Event that is the focus of the communications is in the future	• Event that is the focus of the communications is about to occur or is already occurring
• Ongoing process between communicator and audience is time consuming	• Shorter process between organization and audience due to the immediacy of the crisis event
• Focus of efforts is on the dialogue generated between the two parties	• Focus of the efforts is the delivery of messages to the audience
• Most communications are two-way events	• Most communications are one-way events
• Goal is to reach a consensus with audience regarding activities and solutions to presenting hazard	• Goal is to inform and compel the audience to action intended to keep them safe
• SH&E professional functions include assisting in the risk assessment process to qualify and quantify the risks and assisting in the development of the messages; in some organizations, the SH&E professional will also deliver the messages, typically to the workforce	• SH&E professional functions include assisting in the understanding of the severity of the crisis and assisting in the development of the messages; in some organizations the SH&E professional will also deliver the messages, typically to the workforce

significant crisis event is happening or is about to happen (Sandman 2004) (see Table 2.1.)

The critical difference is the situations in which the various communication forms take place. Risk communication is an ongoing process that helps to define a problem and solicit involvement and action before an emergency occurs. It is a time-consuming process that involves developing relationships with audiences, sharing information about the nature of risks, and working toward a consensus about the best ways to approach the risk. As will be discussed in subsequent chapters, the partnership aspect of the risk communication process is critical. The role of the safety, health, and environmental (SH&E) professional in the process will vary from one organization to another and from one risk situation to another. The key functions often include assisting in assessment of the risk through standardized methods (see Chapter 8). This process allows for messages to be crafted based upon quantitative and qualitative risk measurements in addition to the specific goals and objectives of the organization and the message delivery setting (see Chapter 5). Occasionally, the SH&E professional will be asked to deliver the message(s) in formal settings such a press conferences or organizational meetings with the workforce. Sometimes the message delivery is informal, as in conversations that might occur during walk-throughs of the manufacturing floor or construction site.

Crisis communications are those messages that are given to audiences during an emergency event that threatens them either immediately or at some foreseeable point in the near future. Because of the urgency of the situation, the time needed to develop a partnership with audiences and come to consensus on appropriate actions is rarely available. In addition, the time needed to develop goals and objectives for the messages is short. In crisis communications, SH&E professionals often fill roles similar to those in risk communications efforts, but their need to know and understand the crisis situation, the audience, and the goals of the communication effort is increased. Their value to their organizations in the process is enhanced if they are suitably prepared to act swiftly and provide relevant advice.

THE STAGES OF A CRISIS

Prior to moving on from the definitions of the various terms used throughout this text, some commonly accepted frameworks for identifying crisis stages will be presented. This will aid in developing a stronger understanding of how the development of a crisis affects how messages are crafted and eventually delivered.

Numerous theories and frameworks have been promoted over the years to address the life cycle of a crisis to aid in understanding the overall management of a crisis and the actions taken by an organization in each stage. One of the earliest and most useful for understanding crisis communications was developed by Fink and published in 1986. Fink's four stages of a crisis include the following:

1. **Prodromal.** An organization is able to identify clues or hints that a crisis is about to occur. Not all organizations have readily developed detection systems in place for identifying prodromes and might argue that a crisis occurred without any warning, but most crisis experts agree that there are very few crises that do not provide clues if an organization is watching and looking for them.
2. **Crisis breakout.** This stage represents the earliest events that are part of the crisis and begins with an identifiable event that produces a specific type of damage, physical or reputational, to the organization.
3. **Chronic.** This stage occurs as the organization attempts to address the effects of the crisis. The length of this stage varies greatly and the activities of an organization serve to either reduce the length and effect of the crisis, or, unfortunately, extend it and spread the damaging effects even further. The fundamental principles of crisis management, while not the topic of this text, describe what an organization should do in this stage.
4. **Resolution.** At some point in the crisis, the organization determines that the events are no longer problematic and no longer affect the various audiences. The effects of the crisis may still linger for an extended period of time, but the immediate and most dangerous effects have been resolved, either on their own or due to the actions of the organization.

While significantly more information will be devoted to crafting crisis communications messages in future chapters, it is helpful to close this section with a few comments on the overall framework of messages that are useful based upon the following stages.

Ideas about how to manage crisis in the various stages are, once again, not the topic of this text, but message development and delivery are. Several years after Fink's publication of the crisis stages noted above, Sturges (1994) began to promote the ideas of various messages during different crisis stages. He suggested that audiences are most receptive to basic information during the acute stages of a crisis immediately after the occurrence of the triggering event and shortly after. The focus of the content should be around what is happening, how it will affect the audience, and what they can do to protect themselves. As will be discussed in later chapters, crisis communication messages are more often simple and one-way. Information is delivered as quickly and succinctly as possible. During the resolution stage, messages can be more detailed and feature two-way dialogues between the organization and the audience. Message goals in this phase tend to be focus around what can keep the crisis from occurring again and how the organization can reestablish a positive relationship with the audience.

THE PROCESS OF COMMUNICATION

A full understanding of risk and crisis communications is aided by a brief discussion of the process of communication (Fig. 2.2). The generic steps of nearly every verbal and written communication involve the following (Berlo 1960):

1. A message is sent by a communicator.
2. The message is received by the receiver.
3. The receiver interprets the message.
4. The receiver sends feedback to the communicator about the message.

These steps are repeated as often as necessary and become a circular process that continues until the communication event is finished. The communication event can range from one simple message sent and received to a lengthy "conversation" of hundreds of different messages that go back and forth between the original sender and the original receiver. In such scenarios, the roles of sender and receiver switch back and forth depending upon the originator of the message. Any message in a communication event can be adjusted based on the feedback from the audience, and sometimes a new message is sent on the same topic as the original message, on a separate subtopic or an entirely new topic. In addition, some communication events involve more than one receiver and more than one sender, as in a group discussion or a presentation delivered by one person to more than one person. Obviously these events involve multiple messages being sent and received, often at the same time. However, regardless of the complexity of the individual communication event, the simple model illustrated above still provides an accurate representation of the process.

Finally, messages received are often encoded to determine their perceived meaning by the receiver, which is not always the intended content of the sender. Feedback messages are often sent back and forth between the participants in the communication event and become a sublevel of messages, if you will. Messages are often filtered or changed

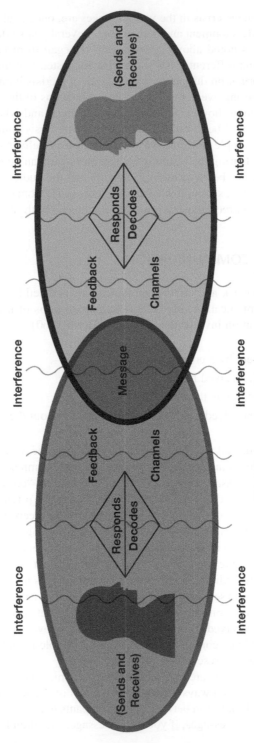

Fig. 2.2. Illustration of the Communication Process

by the receiver based upon their own backgrounds, experiences, and opinions, along with their educational levels and emotional states. These filters are often called interferences and vary among individuals and groups. The message sender needs to take the time to evaluate and understand the receiver's characteristics and emotional state to increase the chances of successful receipt of a message. (Audience profiling is addressed in greater detail in Chapter 4.)

Although many view communication events as an exchange of verbal messages, the communication cycle utilizes both oral and written methods. In traditional risk and crisis communication efforts, the oral methods include press conferences and briefings as well as safety meetings. Safety training is also a form of oral risk communication, as the process typically involves an initial educational component that attempts to help the workforce understand the risk posed by unsafe behavior or the result of not following safe job procedures. The second part of a typical training session more often involves teaching the workforce how to perform a job safely. Risk communication in the form of safety training can occur in the more traditional classroom settings with the use of lectures videos, or it may happen through the use of group and individual exercises. However, more informal safety trainings often called "tailgate meetings" and "toolbox talks" also meet the criteria of a basic risk communication event. Even the more recent forms or oral communications such as audio conferences, podcasts, and robocalls can sometimes take the form of risk and crisis communication events.

Written methods of risk and crisis communications include press releases, brochures, policies and procedures, and newsletters. More recently, organizations have increased their use of e-mail, blogs, and company websites as means of providing risk and crisis communications.

THE PURPOSE AND OBJECTIVES OF THE COMMUNICATION EVENT

As noted above and discussed in Chapter 4, success in risk and crisis communication efforts is increased by understanding the audience and tailoring messages to their needs. Before that process can begin, however, the organization needs to develop a clear understanding of why it is communicating, the purpose and what it hopes to achieve through the effort, and its objectives. Objectives often also answer the "how" questions of the specific types of communication efforts in addition to the frequency and the content of the messages.

One method of understanding the purpose and objectives of the communication effort is to consider what the communicator wants the audience to do with the messages. Sandman suggests that there are three basic scenarios for risk communications. His theories are discussed in greater detail in Chapter 3, but a brief summary at this juncture will assist in clarifying the key process of understanding the purpose and objectives of the communication effort (see Fig. 2.3):

Scenario #1. There is a significant hazard situation, but the audience is uniformed, apathetic, or both, in varying degrees. Therefore, the purpose of the communication event is to increase the audience's understanding of the risk or crisis

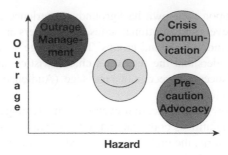

Fig. 2.3. Three types of risk communications—Peter Sandman

that affects it. The objectives of the messages are to persuade the audience members to be more concerned than they currently are and to take a specific action. This type of communication effort involves messages that provide details in a simplified level and are repeated often enough until the audience not only begins to change its opinion on the hazard, but also agrees with the recommended action and performs it. Sandman calls this "precaution advocacy" (Sandman 2007).

Scenario #2. The audience is angry, upset, or excessively worried, but the actual risks of the hazard, as quantified by a formal risk assessment; do not justify the audience's level of concern. Sandman calls this "outrage management," and the purpose of the messages is to both reduce the excessive emotion of the audience and to help it form a more realistic understanding of the risk of the hazard. The objectives of outrage management messages require a series of communication events that focus on acknowledging both the legitimacy of the audience's fears and concerns, as well as acknowledging the role, or even mistakes, of the organization in creating the situation at hand. These messages are then followed by recommendations for more realistic actions that the audience can take (Sandman 2008).

Scenario #3. The audience is fearful and upset about a situation that is serious and is either already occurring or is about to occur. Sandman calls this "crisis communications," and it is in sync with the definition used above to differentiate these types of communication events from risk communications. The purpose of these communication events is to clearly articulate to the audience what is happening and what it needs to know about the crisis in order to act appropriately. Objectives include adjusting the audience's emotional level with messages that do not downplay the risk so that audience members are able to act in their own best interest to protect themselves, yet not panic (Sandman 2004).

The opposite perspective of what is discussed above is not what the communicator wants the audience to hear, but what the audience wants to hear from the communicator. Morgan *et al.* (2002) of Carnegie Mellon University of Pennsylvania offer the following view of the various goals of the communication effort from the audience's view:

- The audience wants **advice and answers**. The members need the kind of detailed information they do not have the time and knowledge to find on their own. They also need instructions on what to do in the form of options that follow their own social forces as well as those of their communities and/or how they might develop their own options. In this scenario, the audience relies on the communicators to be trusted experts who provide the best technical knowledge in a manner that is devoid of vested interests. Communicators preparing messages for audiences will need to provide comprehensive information that includes substantive details.

- The audience wants **information** so its members can make their own decisions about what to do. They want to make their own choices and rely on technical experts to provide them with the information they need that can be evaluated within the framework of their value system and personal situation. Communicators in these situations should first look at the decisions the audience members need to make (or says they need to make) and then work backwards to determine what information is most valid for the decisions at hand. These communication events often involve smaller amounts of information—key points that summarize the situation and what is known about it.

- The audience wants the **processes and a framework** for understanding the situation at hand. These communication events assume an active audience that wants to be involved in helping to assess the risk, identify risky situations, and help develop an action plan. Communicators in these events need to be prepared to fully invest in the interactive process described in the definitions for risk communication noted above and needs to be willing to treat the audience as partners in the communication effort rather than recipients of a message.

While it is all well and good to start any communication effort with a full understanding of the purpose and objectives of the messages, numerous factors should influence the process. Lundgren and McMakin (2004) identify four significant factors:

1. **Legal issues and regulatory requirements.** Many such laws involve the environmental arena as noted at the beginning of this chapter (CERCLA, SARA); however, OSHA and other federal and state regulatory agencies have policies on the requirements of the efforts. For example, OSHA's requirements under the hazard communication standard delineate an organization's responsibility for evaluating and communicating risks to workers from hazardous substances through the use of labels (OSHA 1996). In Pennsylvania (as is true in many other states), the Pennsylvania Community and Worker Right to Know Act adds to these requirements by mandating an annual chemical inventory of hazardous substances and communicating that list by posting it in a location frequented by the workforce (Pennsylvania Department of Labor 1994).

2. **Organizational requirements.** Many organizations have specific written policies for communicating with stakeholders as well as paid professional staff whose responsibilities include both crafting and delivering the messages (public relations departments, human resources departments, and public information

officers). Their advice and permission is often required before any risk or crisis communication effort can occur.

3. **The risk itself.** As noted above and discussed in Chapters 3 and 8, quantifying the risk of a part of the communication process so that the specific content of the messages can be crafted. The risk will also determine whether or not an organization is dealing with risk communications, when time is available to develop partnerships and create interactive processes with audiences, or crisis communications, when the goal may be simpler and focuses on rapid information flow.

Along with understanding the factors that influence the development of communication efforts, constraints also play a role. Risk and crisis communicators need to be cognizant of a variety of factors that impact that goals and objectives of the communication effort (Lundgren and McMakin 2004). First and foremost are the constraints of the organization itself. Management will need to be convinced of the necessity of the communication effort, and their reactions range from enthusiastic support to apathy or hostility. Resources may or may not be available to the communicators to achieve their goals. Organizations may have complicated and time-consuming review and approval processes that delay delivery of the message or end up stopping it entirely. Some of the review processes are designed to protect the organization and some are the result of conflicts among various departments and workgroups about how to manage a risk or crisis communication effort.

Emotional constraints are those which exist within the communicators themselves. Ideally the communication effort is based upon objective data and facts, but those crafting and delivering the messages are human beings acting on a full range of human emotions and with a broad continuum of expectations, realistic or not. One of the most common constraints is the inability of the communicators to see the public as an equal and legitimate partner, a critical function in risk communications where the process of developing partnerships and decisions based upon consensus can significantly impact the success of the effort. Sometimes this perception is based on a belief that an organization is responsible to make decisions about risks and how to manage them, and sometimes this perception is based on the notion that audiences are not capable of fully understanding the high degree of science behind risk assessment. This leads to messages that can be patronizing or "dumb downed" in such a manner that the typical audience response is one of anger, frustration, or hostility. Finally, individual values can also impede the process for both the communicators and the audience. Because values are often a unique part of individuals, crafting messages that address the full range of values can be tricky at best.

A secondary level of constraint is imposed by the audience. This critical issue will appear in many other chapters within this text and is only introduced at this time. Audiences have many levels of emotionality when faced with a risk or a significant crisis event, and many of them involve highly charged states such as anger, hostility, fear, dread, or frustration, all of which exist on a continuum based upon the event, the audience, and a multitude of other factors. (See Covello's risk perception theory and Sandman's Risk = Hazard + Outrage paradigm, both in Chapter 3).

In addition to the emotionality of the audience, each communication event is characterized by varying levels of knowledge and expertise on the part of the audience, not to mention education levels, cultural barriers, and other significant audience characteristics. Messages will be more likely to be heard, understood, and acted upon when they are crafted to the varying levels of audience understanding in addition to emotional states.

REFERENCES

ANSI/AIHA Z10-2005. 2005. *Occupational Health and Safety Management Systems Standard.* Fairfax, VA: American Industrial Hygiene Association.

Berlo, D.K. 1960. *The Process of Communication.* New York: Holt, Rinehart & Winston.

Coombs, W.T. 1999. *Ongoing Crisis Communications: Planning, Managing, and Responding.* Thousand Oaks, CA: Sage Publications, Inc.

Covello, V. 2008. "Risk Communication: Principles, Tools and Techniques," posted online at http://www.maqweb.org/techbriefs/tb49riskcomm.shtml on February 25, 2008. Accessed on October 29, 2009.

Fearn-Banks, K. 2007. *Crisis Communications: A Casebook Approach*, 3rd ed. Mahwah, NJ: Lawrence Erlbaum Associates.

Fields, J. 2008. "Risk Assessment and Hazard Control: Regulatory Issues." In *The Safety Professionals Handbook: Technical Applications*, edited by J.M. Haight, 19–36. Des Plaines, IL: The American Society of Safety Engineers.

Fink, S. 1986. *Crisis Management: Planning for the Inevitable.* New York: AMACOM.

Lundgren, R.E. and A.H. McMakin. 2004. *Risk Communication: A Handbook for Communicating Environmental, Safety, and Health Risks*, 3rd ed. Columbus, OH: Battelle Press.

Morgan, M.G., B. Fishhoff, A. Bostrom, and C.J. Atman. 2002. *Risk Communication: A Mental Models Approach.* New York: Cambridge University Press.

Occupational Safety and Health Administration. 1996. *Hazard Communication.* 29 Code of Federal Regulations, Part 1910.1200.

Pennsylvania Department of Labor and Industry. *The Worker and Community Right to Know Act.* Public Law 734, No. 159 (1994).

Peters, R.G., V.T. Covello, and D.B. McCallum. 1997. "The Determinants of Trust and Credibility in Environmental Risk Communication: An Empirical Study." *Risk Analysis* 17(1):43–54.

Presidential/Congressional Commission on Risk Assessment and Risk Management. 1997. *Framework for Environmental Health Risk Management, Final Report, Vol. 1.* Washington, DC.

Sandman, P. 2004. "Crisis Communications: A Very Quick Introduction." Posted online at http://www.petersandman.com/col/crisis.htm on April 15, 2004. Accessed on January 3, 2010.

Sandman, P. 2007. "Watch Out: How to Warn Apathetic People." Posted online at http://www.petersandman.com/col/watchout.htm on November 9, 2007. Accessed on January 4, 2010.

Sandman, P. 2008. "Managing Justified Outrage: Outrage Management When Your Opponents are Substantively Right." Posted online at http://www.petersandman.com/col/justified.htm on November 19, 2008. Accessed on January 4, 2010.

Sandman, P. 2009. "Dr. Peter M. Sandman: Biography," updated July 2009. Posted online at http://www.petersandmand.com. Accessed January 3, 2010.

Sturges, D.L. 1994. "Communicating Through Crisis: A strategy for Organizational Survival." *Management Communication Quarterly* 7(3):297–316.

U.S. Department of Health and Human Services. 2006. "Communicating in a Crisis: Risk Communication Guidelines for Public Officials." Washington, D.C.

U.S. Environmental Protection Agency. "William D. Ruckelshaus: First Term." Posted online at http://www.epa.gov/history/admin/agency/ruckelshaus.htm. Accessed on December 28, 2009.

3

COMMUNICATION FUNDAMENTALS AND THEORETICAL FOUNDATIONS

This chapter will provide a substantial expansion of some of the basic concepts introduced in Chapter 2. The beginning portion of the chapter will consider the audience factors that can assist in the success of the risk or crisis communication effort and the latter will provide detailed examinations of several of the significant and widely accepted theoretical foundations for risk and crisis communications theories.

AUDIENCE PERCEPTIONS OF THE COMMUNICATOR

The success (or failure) of any risk or crisis communications event is considered to be closely linked in part with the audience's perception of the communicator. Nearly all practitioners and researchers in the field of risk and crisis communications view two key variables as fundamental factors—trust and credibility. The terms overlap in some respects but are very different in others. Understanding how messages are impacted by the levels of an audience as a group as well as by individual audience members is thought by many to be a key to a successful communications event.

Results from early social science research can be used to evaluate some risk and crisis communication efforts, particularly with regard to trust. George Cvetkovich and

Risk and Crisis Communications: Methods and Messages, First Edition. Pamela (Ferrante) Walaski.
© 2011 John Wiley & Sons, Inc. Published 2011 by John Wiley & Sons, Inc.

Tim Earle have conducted studies that look closely at variables of trust as they relate to the establishment of trust by a communicator with an audience. The asymmetry principle suggests that creating trust in an audience by an organization is a difficult task, but once it has been created, positive information about the organization will tend to strongly reinforce an audience's level of trust. Furthermore, reducing or negating the previously established level of trust does not easily occur, even in the face of information to the contrary or if an organization errs in some obvious and public manner. On the other hand, if there is no previously established level of trust or if the level is weak, negative information can easily serve to create and reinforce a level of mistrust about the organization. This audience perception holds true even in the face of contradicting positive information (Cvetkovich *et al.* 2002). Cvetkovich and Winter have also demonstrated that when an audience has limited or no personal control over the specific risk, trust in the organization is a major factor in the audience's acceptance of the risk communications event (Cvetkovich and Winter 2001). [See also Covello's risk perception model (Fig. 3.1).]

Several theories propose and have been tested that tie the audience's acceptance of the message to whether or not the communicator can be trusted to provide accurate information (Peters *et al.* 1997; Renn and Levin 1991). Kasperson has also looked more specifically at the perception of trust, which is a result of the audience's understating of the competence of the communicator, the absence of bias, and of caring and commitment (Kasperson *et al.* 1992; Kasperson 1986). Relevant to the discussion from Chapter 2, the interactive process between the communicator and the audience is a

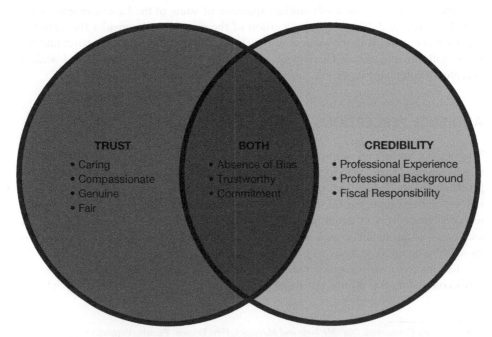

Fig. 3.1. Trust and Credibility Factors in Organizations

fundamental part of risk communications, a process that takes time to develop and mature. On occasion, trust (or perhaps more accurately mistrust) can come from the communicator's representative organization. Certain organizations are viewed by the audience in general to be more trustworthy (religious groups) and other less trustworthy (government groups and politicians).

A second crucial characteristic of the communicator cited regularly in the literature is whether or not the communicator is credible and has the necessary background and experience to know and understand the information conveyed in the message, thereby making the message believable (Peters *et al.* 1997; Renn and Levin 1991). Credibility is somewhat akin to trust, but the difference lies within the audience belief that the communicator is knowledgeable enough about the topic to understand the content.

In addition to the view of the credibility of the communicator, the term can also be viewed as organizational credibility, which is further differentiated by Coombs (1999) into initial credibility, derived credibility, and terminal credibility.

Some organizations, by virtue of their standing in the community or among a particular audience, have a certain amount of positive initial credibility before they even broadcast the first risk or crisis communication message. In addition, by extension, individual communicators as representatives of such organizations assume the same level of positive initial credibility. It should also be noted that the converse is also true with regard to organizations that have negative credibility; their representative communicators will also have negative initial credibility. As has been noted previously, attempting to develop a level of credibility with the audience at the height of a crisis through crisis communications is a difficult task at best.

Once an organization begins to deliver messages, its credibility is derived through the content of the message and the delivery. Overcoming negative credibility and/or developing positive credibility when none previously existed are the main goals of early risk and crisis communication efforts. These efforts require a defined set of goals and objectives as well as effective presentation of well-crafted messages.

Terminal credibility is the credibility that comes after the message(s) are delivered and represents the result of both initial and derived credibility so that the combination of the two variables results in a multitude of outcomes. Positive initial credibility and positive derived credibility result in the strongest level of positive terminal credibility. Positive initial credibility and negative derived credibility results in negative terminal credibility, but of varying strengths. Negative initial credibility and positive derived credibility result in positive terminal credibility, also of varying strengths. And finally, negative initial credibility and negative derived credibility results in strong negative terminal credibility (see Fig. 3.2).

TRUST AND CREDIBILITY

In 1997, Peters, Covello, and McCallum tried to answer the question of what factors could reliably be used to predict an audience's perception of trust and credibility (Peters *et al.* 1997). Although the theories that trust and credibility were significant had been circulating in the literature for some time, they had not yet been empirically tested, and

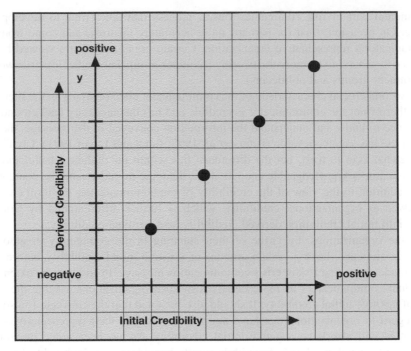

Fig. 3.2. Terminal Credibility

the variables had not yet been evaluated to determine their real-life applicability for risk and crisis communicators. The study's authors believed the details of how to establish the partnership between the communicating organization and different audiences would lead to an increase in the success of the communication event. In other words, if it could be determined *what* increased the audience's level of trust and credibility of the organization, messages and activities could be more specifically targeted toward those goals.

Six unique hypotheses were tested in the study; the primary one dealt with perceptions of trust and credibility and suggested that they were dependent upon three factors: (1) perceptions of knowledge and expertise, (2) perceptions of openness and honesty, and (3) perceptions of concern and care.

The study's methods included telephone surveys of members of the general public selected by random-digit dialing and using a four-point, Likert-type scale. The focus of the study was on risk communication events involving hazardous materials incidents. Respondents were specifically selected from communities where a significant industrial presence existed related to the storage or production of hazardous materials, the existence of a Superfund site, the existence of an active environmental group, and prior emission problems or enforcement activities. Due to the limitations of the populations surveyed, the study's authors were careful to indicate that the results could not be completely extrapolated to all risk communication events but more to those communities with comparable events. Regardless of the study's limitations, however, most risk and crisis communication experts point to it as a landmark event in the understanding of the importance of the two variables and have used the study's main conclusions to

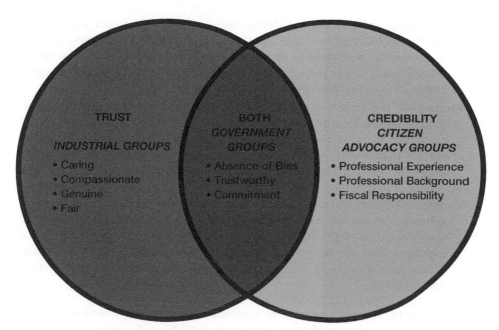

Fig. 3.3. Trust and Credibility in Various Organizations

buttress their arguments about the risk and crisis communications process and the crafting of effective messages.

The findings of the study were differentiated into three organizational groupings (the industrial sector, the government sector, and citizen advocacy groups) as some of the data showed differences in implementation among the three groups. Below are three key findings from the study (see also Fig. 3.3):

Finding #1. In the *industrial sector*, an increase in the audience's *perceptions of concern and care* provides for the largest increase in trust and credibility by the public of the organization. A common stereotype of many industries is the perception that the organization is more concerned about profits rather than people. Therefore an industrial organization that can use risk communication events to develop or increase the audience's level of perception that the organization also cares about what happens in the community is likely to be more successful.

Finding #2. In the *government sector*, an increase in audience's *perceptions of commitment* provides for the largest increase in trust and credibility. The common stereotype about governmental organizations is that they lack stability, that the political party in power determines the goals and efforts of the organization, which may not always be what is best for the audience. And when election results change those in control of the organization, the commitment to previous causes or efforts may be moved to a much lower level of priority or even be completely ignored. The notion that most politicians are looking out for themselves and their ongoing electability is one that can be seen in voter polls and "letters to the editors," among other similar venues. Crafting messages that overcome this perception of the audience and demonstrate a sincere ability to commit to an effort or project over the long haul are likely to generate the largest change in audience perceptions of trust and credibility.

Finding #3. In dealing with *citizen advocacy groups*, an increase in *public perceptions of knowledge and expertise* provides for the largest increase in trust and credibility. This finding is intuitive as most members of citizen advisory groups are not traditionally professionals in the field. They tend to be community leaders or individual community members who have a particular passion for a subject or issue, so their commitment is generally accepted even if not everyone agrees with their opinion. And even if members of the audience hold diverse or contrary opinions, the audience tends to accept their level of concern and care. Communication events that are designed to demonstrate to an audience that the communicators who have a firm grasp of the technicalities of the various issues are the ones most likely to succeed at increasing the audience's level of trust and credibility.

According to the study's authors, the common theme in the above findings was the perceived stereotypes about each type of organization by the study's respondents, which provides a glimpse into some specific and pragmatic tasks and activities that can be undertaken to increase the success of any risk communication effort by any one of the three groups. In general, the study's authors argue that defying negative stereotypes can be significant and necessary if the audience's existing perceptions of trust and credibility are to be overcome so that the risk communications messages are heard, understood, generally agreed to, and, if necessary, acted upon.

The concept of defying stereotypes is one also supported by Sandman in many of his writings as well as those of Fearn-Banks. In her analysis, Fearn-Banks (2007) considers the actions of Johnson & Johnson following the tampering of its Tylenol pain reliever to be a "textbook" case of successful crisis communications. At the beginning of the crisis, Johnson & Johnson enjoyed a positive relationship with consumers who purchased their products, as well as with their employees and the media, three different but critical audiences for the delivery of crisis messages. Despite having never dealt with a similar situation with which it could have prepared a crisis communications plan in advance (in fact, no major consumer product manufacturer had ever dealt with a similar situation that Johnson & Johnson might have learned from or used as a prodrome), the company reacted swiftly and decisively to recall the product and demonstrate *caring and concern* for its consumers well above the financial losses to the company—people over profits, as opposed to what would typically have been expected by the audience. Only when incontrovertible evidence surfaced that the tampering was done by an outsider did Johnson & Johnson attempt to shift the responsibility for the event externally, while still taking critical steps to protect consumers. In the end, while the crisis cost the company over $100 million in sales and other losses, it was able to quickly regain its entire market share in a fairly short period of time. To this day Tylenol remains a highly popular and profitable product.

Sandman also uses the Tylenol tampering scandal of 1992 as the classic example of an organization successfully combating the negative stereotype of profits over people. His thoughts about defying stereotypes appear in numerous articles and website columns. One particular article that addresses this concept, along with the addition of the audience's perception of caring and concern, is titled "Empathy in Risk Communication" (Sandman 2007a). In it Sandman suggests that showing empathy often involves acting in a "profoundly" counterintuitive manner. The concepts and methods of implementation presented in this article focus more heavily on crisis communications, but they can also be applied when the audience is highly upset or overly concerned about a hazard that is relatively low risk. More about the latter concept appears in Chapter 5.

Sandman's writings on this topic postulate that reassuring the audience is a fundamental part of any crisis communication message as well as many risk communication messages. Part of the reassurance involves first acknowledging the audience's fears, worries, and concerns, and communicating that the emotions are shared if the situation calls for it. He references the now-famous quoted answer by Mayor Rudolph Giuliani to a reporter's question about the number of casualties expected in the early hours of the 9/11 tragedy. Giuliani said, "The number of casualties will be more than any of us can bear ultimately" (Giuliani 2001). While many suggest that it was Giuliani's assertive leadership that was essential to dealing with the tragedy in the first hours, weeks, and months, Sandman insists that much of it would not have been possible if Giuliani had not first established the fundamental connection between himself as a fellow human being grappling with the same feelings of anguish, fear, and anxiety as his audience. Sandman strongly cautions crisis managers who believe that crisis communications are most effective when audiences are told not to worry and that the organization has everything under control when he says: "Crisis managers who imagine that showing empathy means over-reassuring people, 'emphasizing the positive' or 'calming them down,' are way off the mark" (Sandman 2007a).

Sandman also addresses what to do if, as a communicator, empathy is not an intuitive behavior. This may occur because of the personality of the communicator and may need to be developed through time, practice, and even "faking it" for a time. In addition, it is sometimes the result of simply not understanding the feelings of the audience, despite a desire to do so. He suggests adopting the attitude of trying to understand, allowing the basic level of caring and concern to come through in the messages, helping to audience to realize that the communicator is making a sincere effort, even if the ability to understand the feelings of the audience isn't there at the moment. When all else fails, finding a different communicator is essential if the communication events are to be successful.

FOUR THEORETICAL MODELS

Covello and his colleagues at The Center for Risk Communication have published their thoughts on how risk information is processed, how perceptions are formed, and how decisions are made by members of the audience. The four models described below provide a working foundation to enable risk communicators to successfully craft messages (Covello 2007). These models can apply in risk communication events that are more long term and involve processes between communicators and audiences as well as those crisis communication events that often have to be developed and delivered in fairly short periods of time. As has been discussed earlier in this chapter, understanding audience perceptions are fundamental to successful message delivery.

One additional comment prior to describing the four models is to note that concepts of one model may overlap between models. It should be noted that a full understanding of the theoretical foundations of risk and crisis communications, at least from Covello's perspective, is that there is no one simple and straightforward theory to encompass all situations and that communicators should provide a clear analysis of each and every audience. The usefulness of the models below is that they provide a variety of theories and ideas that can be applied as needed, depending upon the presenting risk or crisis event.

The Risk Perception Model

This model identifies factors that influence an audience's perception of risk and provides for an analysis of the magnitude of the perception by the organization doing an audience profile. Covello and colleagues use this model to elucidate 15 of the factors they believe most important in the analysis of an audience due to the critical role they play in analyzing audience levels of concern and other strong emotions such as fear, worry, hostility, and outrage. Understanding these factors and using them to then profile an audience helps to craft messages more likely to achieve their stated purpose and objectives by changing attitudes and behavior. (The latter factors are reflected further in discussions below with regard to Sandman's paradigm.)

Table 3.1 discusses these factors in a positive vein; it is critical to note that the opposite of each also holds true. Several additional key points are also necessary for understanding these factors. The first is that each unique risk or crisis situation will produce its own unique combination of the 15 factors so that an inestimable variation is possible. Some factors will be more prominent in the analysis, some less so, and some will not apply at all. In a similar vein, the intensity of each factor will vary for each situation so that some factors may figure significantly in the analysis and others may have limited or no applicability. In addition, a comprehensive audience analysis may also identify subgroups whose perceptions of the factors will vary so that it may be complicated to develop one unique audience picture. And finally, as a crisis develops, the audience's perceptions of key factors may change, significantly at times, rendering the earlier analysis inappropriate for the current situation.

It might seem as though the large number of factors may make an audience analysis difficult and time consuming at best and at worst nearly impossible. However, the reader is cautioned to utilize the factors as a guide and a reference point, allowing for fluctuations due to the changing nature of both the event and the audience.

Later chapters will delve more deeply into utilizing the above factors to craft and deliver risk and crisis messages; however, one obvious implication of the use of these numerous risk factors is to try and segregate out an audience into subgroups for whom

TABLE 3.1. Covello's 15 Risk Perception Factors[a]

Risk Factor	Applicability
Voluntariness	If the audience members perceive the risk to be voluntary, they are more likely to accept it because they understand their role in experiencing the implications of the risk.
Controllability	If the audience members perceive that they have control over the risk, they are more likely to accept the implications of it.
Familiarity	If the audience members have some previous knowledge of the risk or experience with it, they are more likely to accept the implications of it because of the increased level of knowing what might or might not happen.
Equity	If the audience members perceive the implications and consequences of the risk to be equally shared among audience members, they are more likely to accept the implications of it.
Benefits	If the audience members perceive the ultimate benefits of the risk to be positive, they are more likely to accept the potential negative implications of experiencing it.
Understanding	If the audience members possess a basic understanding of the risk, they are more likely to accept the implications of it. The greater the level of understanding, the higher the acceptance.
Uncertainty	If the audience members perceive the risks have a degree of certainty in various dimensions and in the scientific information available about it, they are more likely to accept the implications of it.
Dread	If the audience members' emotions with regard to a risk are less intense and fearful, the more likely they are to accept the implications of it.
Trust in institutions	If the audience members perceive the institutions more significantly involved in the risk as trustworthy and credible, the more likely they are to accept the implications of it.
Reversibility	If the audience members perceive the risk to have reversible adverse effects, they are more likely to accept the implications of it.
Personal stake	If the audience members perceive the risk to be limited in its personal implications and consequences, the more likely they are to accept the implications of the risk.
Ethical/moral nature	If the audience members perceive the risk to be morally or ethically acceptable, they are more likely to accept the implications of it.
Human vs. natural origin	If the audience members perceive that the origin of the risk is naturally occurring, they are more likely to accept the implications of it.
Catastrophic Potential	If the audience members perceive that the amount of fatalities, injuries, and illnesses from a risk are minimal, they are more likely to accept the implications of it.

[a]Covello et al. 2001

messages can be uniquely crafted. Some audiences naturally segregate themselves such a workforce and the surrounding community, but within each of the major groups may be many subgroups for which different messages may need to be created to achieve optimum results.

The Mental Noise Model

Any risk or crisis situation produces stress in an audience. The 15 factors noted above can help calibrate that stress level within an audience and between audience groups, but a certain amount of baseline stress is to be expected within the audience when engaging in risk and crisis communication events. Stress produces what Covello and his colleagues call "mental noise": the higher the level of stress and anxiety, the higher the amount of mental noise. Consequently, events that produce a higher level of mental noise within an audience reduce its ability to process information and messages.

The level of mental noise exists on a continuum and is generated from a variety of factors. Covello suggests that the following factors, some of which appear above in the risk perception model, cause the highest levels of fear and worry (Covello 2007):

- The level of *control* of the audience over the outcome and whether or not the audience trusts the other parties and sources of information who may have some or all of the control if the audience does not
- Whether or not the crisis situation is *voluntary* and/or *escapable*
- Whether or not the crisis is *man-made* or a natural disaster
- Whether or not the crisis is not *familiar* to the audience or extremely unusual so that the audience may not have had any experience in the past with which to calibrate its response
- The threat of an illness or injury from the crisis that typically produces *dread* in the audience (e.g., cancer)
- If there is significant *uncertainty* about the situation, its development, and/or outcomes
- If the most likely *victims* of the crisis are seen as helpless, such as children, pregnant women, or the elderly.

Risk and crisis communications need to be able to get beyond (or under or around or through) the mental noise being experienced by the audience if the messages being delivered are to be heard, understood, accepted, and acted upon.

The Negative Dominance Model

In any risk or crisis event, an audience is required to process both positive and negative messages containing information vital both to how they perceive the situation and how they act upon it. This model postulates that situations producing risks and subsequent emotions of fear, anxiety, dread, hostility, or outrage create an environment where an audience is more likely to actually hear and integrate negative messages. This is in part because the negative messages support an audience's "negative" emotions and also because psychological theory would suggest that people are often more focused on negative outcomes rather than positive ones—the classic "grass is greener" concept.

This model further identifies two practical implications for crafting messages: one is that an audience is more likely to hear positive messages if they "overbalance" the negative ones or occur at a greater frequency. Positive messages are those that either assist the audience in moderating the danger or implications to themselves, or provide an action that the audience can take to increase their perception of some of the factors noted above, such as controllability, voluntariness, and benefits. The second is that messages containing negative words such a "never," "no," "not," "nothing," and "none" are more likely to be remembered by the audience and for longer periods of time. They also may create a greater impact than positive messages, which should focus on what is being done to mitigate the event and protect the audience. Risk and crisis communicators may need to practice removing such negative terms from their vocabulary when engaged in message communication events.

The Trust Determination Model

The criticality of establishing and maintaining trust between the communicator and audience has been elaborated upon above and will also provide a central theme throughout this text. This model addresses this concept and summarizes the results of the research study discussed above.

Fundamental to the establishment of trust is an understanding of the time commitment on the part of the communicator. Actions that are taken in the immediacy of a crisis are not likely to build trust while the crisis occurs, although they may create a level of trust after the event has resolved itself, when the audience is secure and able to take a more distant view of the event. Communicators should accept that they need to identify risks to an audience and begin to build trust through their actions as well as build consensus on both the level of risk and mitigation strategies. This process takes time, requiring actions that demonstrate reliability, credibility, good listening, and good communication skills. As noted above, the trust determination factors that play the greatest role in this model include caring and empathy, competence and expertise, and honesty and openness.

This model further encompasses the concepts of the communicator being a member of a trusted or reliable group as opposed to one generally perceived to be untrustworthy. Trusted groups typically include religious organizations and advocacy groups (when the audience member is a member or supporter of that advocacy group). Groups that generate less trust among the general populations include political groups, government bureaucracies, and large corporations. A communicator who is member of a trusted group possesses an advantage when communication barriers exist or when the audience's emotions are running high. Communicators from traditionally untrustworthy groups have that initial barrier to overcome before the message will be heard, believed, and acted upon.

Finally this model addresses risk and crisis events when more than one communicator is involved. It suggests that disagreements among the various communicators will increase the level of mistrust as will lack of coordination among organizations tasked with managing the risk. An example of this type of scenario can be seen in a reflective view of the crisis communication efforts during Hurricane Katrina in August 2005. As the storm approached the Gulf Coast, it provided the potential to be one of the most serious storms to hit the area in many years (a potential that was sadly realized). The dangerous conditions that resulted from Hurricane Katrina coming ashore in New Orleans and the surrounding communities in August 2005 were substantial, but the lack of coordination among governmental authorities and other relief organizations has provided a more lasting image with regard to trust and credibility as time goes on. In televised news footage and press conferences, interviews with victims of the disaster, and in print reports, viewers observed citizens whose lives were in danger seemingly unable to be rescued, along with critical shortages of water, food, and medicine. News accounts of people perishing as a result were devastating to the observing audience.

RISK = HAZARD + OUTRAGE

Peter Sandman is just one of many risk and crisis communications experts; however, his writings and teachings have dominated the practice since his initial experience at the nuclear power generator plant incident at Three Mile Island in March 1979 (Sandman 2009). From his body of work, the most quoted concept is the paradigm "Risk = Hazard + Outrage," which connects the two variables of hazard and outrage to assist risk and crisis communicators in understanding their audiences as well as the hazards they face. This understanding assists with crafting messages that are more likely to be successful. Both variables exist on a continuum, and determining where they lie in a given situation is the fundamental key to knowing how the audience is feeling and what types of messages will cause perception changes or actions sought by the communicator.

Risk, in this paradigm, is viewed more as a personal perception and not necessarily in the technical quantification methods of risk assessment. While many methods of quantifying risk exist and performing standard risk assessment activities should not be set aside when crafting risk and crisis communication messages to determine actions to take in a crisis event, the messages delivered to audiences need to be based upon the audience's personal perceptions of the situation, not necessarily on a risk assessor's data points. This concept is supported by Covello's risk perception model above and requires a view of risk and crisis communications that is more fluid and based upon what the audience *believes* to be true, even in the face of clear evidence to the contrary.

Sandman (2003) has this to say about the two variables:

In a nutshell, "hazard" is the technical component of risk, the product of probability and magnitude. "Outrage" is the nontechnical component, an amalgam of voluntariness, control, responsiveness, trust, dread, etc. They are connected by the fact that outrage is the principal determinant of perceived hazard. When people are upset, they

TABLE 3.2. Sandman's Four Kinds of Risk Communications[a]

Scenario Variables	Key Concept	Key Phrase
High hazard Low outrage	Precaution advocacy	"Watch out!"
Low hazard High outrage	Outrage management	"Calm down!"
High hazard High outrage	Crisis communications	"We'll get through this together."
Medium hazard Medium outrage	Stakeholder relations	Varies

[a]Sandman (2003)

tend to think they are endangered; when they're not upset, they tend to think they're not endangered.

From these two variables, Sandman has postulated four kids of risk communications, as summarized in Table 3.2.

High Hazard/Low Outrage ("Watch out!")

Sandman calls activities that fall under this scenario "precaution advocacy" (Sandman 2007b). This situation features a serious hazard but an apathetic audience. In these types of situations an audience does not often object to the message and is mostly receptive to the content. The apathy of the audience increases the likelihood that they will listen to most communicators and messages without reservations or objections. However, even with a skilled communicator and message, *changing* the audience's perception of the risk or *moving* them to a desired action or behavior change can be difficult. And in the case of precaution advocacy, the objective is often to have the audience's perception of the hazard match the actual hazard or at least move it further in that direction.

The unfortunate tendency for many communicators in this situation is to exaggerate the hazard scenario in order to "scare" the audience into action. Sometimes this can be effective, but it can also be risky and cause an overreaction by the audience, followed by mistrust when the true nature of the hazard is discovered. (A more detailed discussion and recommended approach to this situation, "worst-case scenarios," is addressed in Chapter 9.)

The task for the risk and crisis communicator is to find the means to convey the message that will predispose the audience toward desired goals. Messages should be short and aimed at increasing the audience's outrage so that it is more in line with the actual hazard and so that it provokes action or at least attention. An example of this type of scenario can be seen by reviewing the Hurricane Katrina crisis. Although many residents heeded the calls to evacuate (the message), certain groups were unwilling do so. Some of those who stayed, not because they couldn't leave but because they chose not to, lost their lives because their outrage level was not sufficiently moved to a desired action (evacuation) (Sandman 2005).

Medium Hazard/Medium Outrage (Stakeholder Relations)

This is the easiest communication environment, and the task is to simply provide an open and honest dialogue that explains the situation and allows sufficient opportunity for audience response and questioning. It is likely that the audience will also heed the request for action.

These types of scenarios lend themselves to lengthier processes of dialogue and consensus decision making between the communicator and stakeholders and have been discussed in other sections of the text as risk communications rather than crisis communications. The processes of community engagement stipulated in the Superfund program work well here because the hazard is not immediately life threatening and allows the time necessary to develop a consensus on site hazard and remediation decisions. Further, the audience is easier to engage because the hazards often affect their homes and families (U.S. EPA 2005).

Low Hazard/High Outrage ("Calm down!")

This is the most difficult scenario for a risk communicator, as the audience is often operating on a high level of mistrust of both the organization and the individual communicator. This latter critical factor, which has been discussed above, must be addressed before any message is to be believed by the audience and acted upon. Further complicating this scenario is that audiences are sometimes controlled by a small group of "fanatics" who purposely exaggerate the situation for varying motives. These subgroups may also truly believe that situation is dire when the facts say otherwise or at least suggest that the situation is not nearly as serious as some might believe.

The tasks for the communicator in this scenario are to reduce the outrage by sincere listening, acknowledging, and even apologizing, if that will move the audience to a more realistic view of the seriousness of the hazard. The advantage here is that due to the high level of outrage the communicator does have the audience's attention, and with skillful messages, movement in a desired direction is possible.

An example of this type of scenario occurred just months after the Exxon Valdez disaster in Alaska in 1989 when a BP oil tanker spilled a much smaller amount of oil off the coast of California. Realizing that Exxon erred by not quickly providing timely information to the public about the spill, BP averted a similar public relations nightmare by providing immediate, regular, and timely communications about the spill, the effects, and the clean-up efforts. Even though the spill was substantially smaller than the Valdez incident (low hazard), the outrage level of the residents of California initially began as high due to the events in Alaska. The perception on the part of the audience was that of being lied to and kept in the dark about the realities of the situation by the oil companies. In a relatively short period of time, through skillful messages that demonstrated concern and action on the part of BP, the audience outrage level was successfully moved closer toward the actual hazard level (Fearn-Banks 2007). (As will be further discussed in Chapter 10, the irony is that BP's actions in California in 1989 were not sufficiently repeated during the Deepwater Horizon oil rig explosion and spill in the Gulf of Mexico in 2010.)

High Hazard/High Outrage ("We'll get through this together.")

Sandman suggests that this scenario is relatively rare but cites the September 11, 2001, terrorism attack in New York City as an excellent example. The audiences in these types of situations are not nearly as angry as they are fearful and scared, and because the hazard is serious, their position may be valid. (However, it should be noted, as was experienced in the 9/11 attack, that once the terror fades and immediacy of the danger passes, anger may be the next emotion an audience generates.) Without skillful management by the communicator, the outrage can easily slip into terror or depression, both of which are of limited use in moving the audience to take the desired action.

The communicator in this situation must tread carefully, allowing for the audience's legitimate fears, while remaining human and empathetic but still rational, and demonstrating true leadership. The advantage for the communicator is that the outrage is not typically directed at them, at least until after the crisis is past.

As a follow-up to the example presented above regarding the days just before Hurricane Katrina struck, the situation quickly deteriorated into an example of this scenario. Media reports showing desperate and dying citizens of a major United States metropolitan city created incredulous emotional states across the country (exceedingly high outrage). Compounding the problem was that early efforts by government officials to rescue those in need and alleviate suffering proved unsuccessful; it was difficult for the audience to understand why the situation was occurring. Fortunately, the appointment of Lt. Gen. Russel Honoré to lead the Joint Task Force and his communication events were exactly the type of crisis communications that were needed at the time. Honoré was often praised for his brash leadership skills, clearly communicating the gravity of the situation and the need for swift action, all the while demonstrating empathy for the citizens of New Orleans (Duke 2005).

MENTAL MODELS

Like Covello at The Center for Risk Communication and Sandman at Rutgers University, a large body of work has been developed by M. Granger Morgan and his colleagues at Carnegie Mellon University (Morgan *et al.* 2002). The mental models approach is a method of developing risk communications that is based upon sound research and practice with a number of different hazards including radon in homes, climate change, HIV/AIDS, and power-frequency fields.

Morgan describes the mental models theory as intellectual in its approach rather than "do it yourself" and suggests that this more complicated method, which relies heavily on natural science and expert reviews of messages that are tested and retested prior to being delivered in a variety of formats, assures greater success in message acceptance and audience action. The message creation process is long and somewhat arduous, not to mention expensive as compared to many other methods of developing risk and crisis messages, but Morgan argues that this method is not only more likely to succeed because of the lengthy discernment process, but also focuses heavily on audience understanding and acceptance of why a particular action ought to be taken or not taken. He asserts that without understanding why an audience comprehends and responds to certain risks first, messages are more likely to be hit or miss.

In the text supporting and explaining this theory and approach, Morgan *et al.* (2002) suggests that simple and obvious risk messages (such as not smoking in bed) are essentially successful because they rely on audience intuition and logic. However, the seemingly underwhelming success of the simple message "Just Say No" from its first usage in the days of former First Lady Nancy Reagan until the present time speaks to a more complicated problem in message crafting and delivery.

Morgan asserts that the mental models approach draws its strengths from sound psychological theories of human behavior and understanding along fundamental concepts related to economics, natural sciences, engineering, and public policy. He summarizes mental models in his preface:

> At its heart are commitments to the scientific facts of risk, the empirical understanding of human behavior, and the need for openness in communication about risk. We sought an approach that would treat diverse problems with a common set of methods and theories, as well as one that would be readily usable by the professionals entrusted with communicating about risks.

The process of developing communication messages through the mental models approach involves five steps (Morgan, *et al.* 2002):

1. **Create an expert model.** A review of current scientific literature is a necessary first step to comprehend the nature and magnitude of the risk. Through the use of an influence diagram, a network of known information is connected and involves information from a variety of diverse experts. This model is reviewed by technical experts in the area of the hazard to develop consensus, continuity, and authoritativeness of the content.

2. **Conduct metal models interviews.** Through the use of open-ended interviews, audience perceptions of the hazard are solicited. Interviews are structured to follow the influence diagrams. The use of open-ended questions ensures that the responses will be narrative and in the audience's own words, even if the responses are factually incorrect; possibly it is more important to the process that they are. The responses undergo an intense analysis to link them to the influence diagram and elicit areas of fact that are not clear for the identified audience.

3. **Conduct structured initial interviews.** Confirmatory questionnaires are developed that capture the responses expressed in the open-ended interviews as well as the influence diagram. These interviews are then conducted among a larger group than that of Step 2 and focus heavily on the eventual intended

audience. The goal is to identify and quantify the prevalence of certain beliefs among the intended audience.

4. **Draft risk communications.** The information gained from both the interviews and questionnaires is utilized to craft risk communication messages designed to inform by filling knowledge gaps and to correct audience misperceptions. The strength of the audience's misperceptions comes from the earlier steps and is the focus of much of the message content. The drafts are reviewed by the experts to assure accuracy.

5. **Evaluate communications.** Target population interviews occur to test the messages, which are then further refined before widespread delivery. One-on-one interviews, focus groups, and questionnaires are the methods used to test and refine the messages.

Due to the length of time required to work through the five steps, as well as the time and resources required, this model's applicability as presented may be more limited to certain risk communication situations or crisis communications situations when the threat is not imminent but the ability to predict its eventual occurrence is sound. However, attempts will be made through the remaining chapters of this text to flesh out certain portions of the process and describe the applicability in situations that are more typical and not reliant upon grants and other major sources of funding, as the mental models approach tends to be.

FUNCTIONAL LINES OF COMMUNICATION

Regina Lundgren and Andrea McMakin address the various forms of risk and crisis communications, and their approach is helpful in laying out various communications situations and clarifying the types of messages and objectives of each. The functional lines they describe include care communications, consensus communications, and crisis communications (Lundgren and McMakin 2004). These functional lines have obvious overlap as is described below, but the unique characteristics of their applications require risk and crisis communicators to utilize differing tactics and communication methods.

Care Communications

These types of communication lines are best used when the hazard is well characterized and accepted by the audience. This is similar to Sandman's stakeholder relations concept; however, Lundgren and McMakin do not address the seriousness of the hazard in this functional line of communication. The situation could very well pose a significant hazard, but as long as the audience is in agreement with the assessment and the associated dangers, the messages are delivered and generally well accepted. Communicators in these lines of communication include those charged with informing the audience about health hazards, such as the American Heart Association, local public health departments, and televised public service announcements warning about the dangers of smoking.

Care communications also involve the work of many safety professionals when they engage in risk communications through standard safety training activities, including traditional training classes as well as the briefer tool box talk and tailgate meeting format. Messages in this line of communication also include safety posters, newsletters, and other forms of written communication designed to educate and advise about workplace risks and appropriate action to minimize them.

In terms of the previous discussions of trust and credibility, as well as the development of ongoing relationships with the audience, the acceptability of this type of communication is due in part to the audience's reception of the communicator because of the development of trust and credibility. These relationships have taken time to develop and often consist of meaningful two-way communication efforts.

Consensus Communications

This functional line of communication describes the efforts to get meaningful cooperation and consensus from differing audience groups, who may or may not be in agreement. It enjoins those with a stake in the management of the risk to get engaged in the process and help to shape the actions that are derived from the groups' efforts. Lundgren and McMakin also use the terms "public engagement," public involvement," and "public participation" to describe this process.

The Superfund program's community involvement efforts (mentioned earlier) are a good example of this type of process (U.S. EPA 2005). In the beginning there may be substantial differences among the various groups regarding the hazard and the level of risk it presents, but the purpose of the process of consensus communication is to bring differing groups together to jointly develop agreement about the hazard and then to come to a consensus about the best ways to mitigate and remediate the hazard. As with care communication, this is a long-term process that succeeds when the audience can develop trust and credibility in the communicators. Until that occurs, consensus on actions is often difficult to achieve.

In addition to developing consensual strategies among the stakeholders in this process, consensus communication can also serve the purpose of conflict resolution and negotiation when disagreement develops or is present from the beginning of the process.

Crisis Communications

The definition for this line of communication is similar to previous discussions regarding definitions in Chapter 2 in that it occurs in the face of danger that is often sudden, even if predictable. Differences of opinion on the hazard level rarely exist or are so minimal that they do not need addressed in most of the messages that are delivered.

Communicators may or may not have developed previous relationships with the audience and those relationships may or may not be positive, but the time for developing messages is short. Therefore, as has been noted above, the level of trust and credibility of the communicator is the predominant factor as to whether or not the messages will be believed and acted upon. Natural disasters, industrial accidents, and widespread outbreaks of disease are example of situations when crisis communications occur. Planning for such types of emergencies and the associated communication efforts would be considered either care or consensus communications, depending upon the situation and the level of audience involvement in the process.

THE EXCELLENCE THEORY

In her book, *Crisis Communications: A Casebook Approach*, Kathleen Fearn-Banks delves into the development of modern-day crisis communications theory. She traces its genesis to theories proposed in the 1980s regarding public relations excellence. Risk and crisis communications messages are developed using similar methods as those used by public relations professionals to influence key publics and stakeholders (Fearn-Banks 2007).

The excellence theory was first developed by Grunig and Hunt in the mid-1980s to address public relations models and how organizations could achieve the type of publicity they desired through four different models that existed along a continuum of the level involvements of the audience. These models influence the development of the message, provide some framework to help understand audiences, and use those understandings to develop and deliver messages (Grunig and Hunt 1984; Grunig 1992). The excellence theory postulates that most traditional approaches to public relations would suggest that all publicity is good publicity, while the less traditional approaches embrace two-way communications and mutual understanding to negotiate, compromise, engage, and create a dialogue with audiences. The relationship of the excellence theory to the above discussions of risk communications, particularly Lundgren and McMakin's consensus communications, is obvious.

Marra expounded on the work of Grunig and his colleagues by peering more closely at the field of crisis public relations and looked for models that would allow for a better understanding of the variables that create effective crisis communications plans. Marra focused his work on the importance of strong relationships with key audiences, which are developed before a crisis occurs and how those relationships are a clear indicator of how an organization can mitigate its financial, emotional, and perceptual damage following the crisis. His writings expound on Grunig's by aligning strong positive relationships between communicators and their audiences with two-way communications rather than asymmetrical ones, which supports earlier discussions in this and previous chapters regarding the need for ongoing dialogue when trying to develop sound risk communications (Marra 1992).

Fearn-Banks utilizes all of the above theoretical foundations and adds to the theory by suggesting that organizations that utilize thorough crisis inventories to anticipate and plan for crises suffer less financial emotional and perceptual damage. Lastly, she postulates that organizations demonstrating a sound level of openness and honesty in their communications suffer less financial, emotional, and perceptual damage than those who do not. The ideas of openness and honesty are recurring themes in this chapter.

THE "STICKINESS" OF MESSAGES

In his widely popular book *The Tipping Point*, Malcolm Gladwell discusses his beliefs about agents of change in society. His discourse in this context relates to the spread of what he terms "social" epidemics of style, jargon, and television shows. He postulates that there are three factors that determine whether or not a social epidemic will "tip": the law of the few, the "stickiness" factor, and the power of context. Of the three, the stickiness factor relates directly to the ability of a message to be retained by an audience. He further argues that the ability to craft "contagious" messages will increase the chances of them being heard through the barrage of messages our current society produces and has interesting implications for risk and crisis communications (Gladwell 2002).

According to Gladwell, crafting a more memorable (contagious) message can be simply a matter of changing the presentation and structuring of the information. In doing so, a communicator can substantially affect the message's impact. As it relates to risk and crisis communications, a message that has significant impact on the audience increases the likelihood that the audience will be motivated to either change their attitudes about the risk or crisis or be moved to act in a desired manner.

In an example to describe this process, Gladwell discusses a battle between two unlikely competitors for the marketing account of a large record company, one a well-funded public relations organization with large client accounts. This organization's reputation hung on its sophisticated advertisements. The other competitor was the record company's longstanding vendor, a much smaller company with fewer resources and experience with large companies. The smaller company proposed that the two marketers be able to develop an advertising campaign that would run for a period of time and that the results of customer reactions to the campaign would be the deciding factor as to who retained the record company's account. In this classic "David vs. Goliath" endeavor, the smaller company succeeded with a series of low-budget commercials that ran on late night television broadcasts featuring a "gold sticker" that customers could look for in a print advertisement, which they could then use to get a free record with their paid order. Gladwell suggests that the smaller company's advertisements were simple but effective because they provided an incentive to the audience to perform an action, thereby making the message stick. In other words, what made the message effective and memorable was that it encouraged and succeeded in making the audience participants in the process; a common theme in discussions from Chapter 2 about risk communications and the importance of a two-way process directly involving the audience. Using Gladwell's terminology, what can make a message "stick" out from all of the other messages audiences receive each day are those that directly involve them in some action or activity for which they receive a benefit.

REFERENCES

Coombs, W.T. 1999. *Ongoing Crisis Communications: Planning, Managing, and Responding.* Thousand Oaks, CA: Sage Publications, Inc.

Covello, V. 2007. "Effective Risk and Crisis Communication During Water Security Emergencies." U.S. Environmental Protection Agency; EPA 600-R07-027.

Covello, V., R. Peters, J. Wojtecki, and R. Hyde. 2001. "Risk Communication, the West Nile Virus Epidemic, and Bioterrorism: Responding to the Communication Challenges Posed by the Intentional or Unintentional Release of a Pathogen in an Urban Setting." *Journal of Urban Health* 78(2):382–391.

Cvetkovich, G., M. Siegrist, R. Murray, and S. Tragesser. 2002. "New Information and Social Trust Asymmetry and Perseverance of Attributions About Hazard Managers." *Risk Analysis* 22(2):359–367.

Cvetkovich, G. and P.L. Winter. 2001. "Social Trust and the Management of Risks to Threatened Endangered Species." Presented at the Annual Meeting of the Society of Risk Analysis, Seattle, Washington, December 2–5.

Duke, L. 2005. "The Category 5 General." *The Washington Post*, September 12.

Fearn-Banks, K. 2007. *Crisis Communications: A Casebook Approach*, 3rd ed. Mahwah, New Jersey: Lawrence Erlbaum Associates.

Gladwell, M. 2002. *The Tipping Point*. New York: Little, Brown and Company.

Giuliani, R. Quote transcribed from a press conference held on September 11, 2001 with Rudolph Giuliani, Mayor of New York City, and George Pataki, Governor of the State of New York. Posted online at http://transcripts.cnn.com/TRANSCRIPTS/0109/11/bn.42.html Accessed on January 11, 2010.

Grunig, J.E., ed. 1992. *Excellence in Public Relations and Communications Management*. New York: Routledge.

Grunig, J.E. and T. Hunt. 1984. *Managing Public Relations*. New York: Holt, Rinehart and Winston.

Kasperson, R.E. 1986. "Six Propositions on Public Participation and Their Relevance for Risk Communication." *Risk Analysis* 6(3):275–281.

Kasperson, R.E., D. Golding, and S. Tuler. 1992. "Social Distrust as a Factor in Sitting Hazardous Facilities and Communicating Risks." *Journal of Social Issues* 48(4):161–187.

Lundgren, R.E. and A.H. McMakin, A.H. 2004. *Risk Communication: A Handbook for Communicating Environmental, Safety, and Health Risks*, 3rd ed. Columbus, OH: Battelle Press.

Marra, F.J. "Crisis Public Relations: A Theoretical Model." PhD dissertation, University of Maryland, College Park, 1992 (unpublished).

Morgan, M.G., B. Fishhoff, A. Bostrom, and C.J. Atman. 2002. *Risk Communication: A Mental Models Approach*. New York: Cambridge University Press.

Peters, R.G., V.T. Covello, and D.B. McCallum. 1997. "The Determinants of Trust and Credibility in Environmental Risk Communication: An Empirical Study." *Risk Analysis* 17(1):43–54.

Renn, O. and D. Levin. 1991. "Credibility and Trust in Risk Communication." In *Communicating Risks to the Public*, edited by R.E. Kasperson and P.J.M. Stallen. Dordrecht, The Netherlands: Kluwer Academic Publishers.

Sandman, P. 2003. "Four Kinds of Risk Communication." Posted online at http://www.petersandman.com/col/4kind-1.htm on April 11, 2003. Accessed on January 4, 2010.

Sandman, P. 2005. "Katrina: Hurricanes, Catastrophes, and Risk Communications." Posted online at http://www.petersandman.com/col/katrina.htm on September 8, 2005. Accessed on January 29, 2010.

Sandman, P. 2007a. "Empathy in Risk Communication." Posted online at http://www. petersandman.com/col/empathy.htm on July 29, 2007. Accessed on January 11, 2010.

Sandman, P. 2007b. "Watch Out: How to Warn Apathetic People." Posted online at http://www. petersandman.com/col/watchout.htm on November 9, 2007. Accessed on January 4, 2010.

Sandman, P. 2009. "Dr. Peter M. Sandman: Biography," updated July 2009. Posted online at http://www.petersandmand.com. Accessed January 3, 2010.

U.S. Environmental Protection Agency. 2005. "Superfund Community Involvement Handbook." EPA 540-K-05-003.

4

CRAFTING RISK AND CRISIS MESSAGES—SETTING GOALS AND OBJECTIVES AND AUDIENCE PROFILING

This chapter will discuss the process of crafting risk and crisis messages for a variety of circumstances and various audience types. It will begin with a discussion of some overarching goals of message development, including further examination of the importance of developing the purpose and objectives for messages. It will also focus on the importance of profiling an audience prior to crafting messages so that the messages can match the audience's needs and desires, thus increasing their effectiveness. Areas of importance common to most messages will be discussed including dealing with uncertainty, anger and other strong emotions; delivering messages that demonstrate empathy on the part of the communicator; and building or rebuilding trust and credibility.

KEY SUCCESSFUL MESSAGE DEVELOPMENT CONCEPTS

There are three key concepts that influence all message development; the first to be discussed will be the constraints on the message in terms of both the communicator and the audience. As noted briefly in Chapter 2, understanding these constraints in advance of the development of any communication effort will assist in developing solid purpose and objectives to guide all of the activity that follows.

The second is in understanding the criticality of closing the gap between what the audience knows and what it needs to know. A detailed audience profile, based upon a thorough analysis of various factors, is necessary, again in advance of the crafting of any message. The work of Morgan and colleagues at Carnegie Mellon University utilizes a mental models approach to achieve this profile. It is based upon crafting carefully detailed influence diagrams that are reviewed by subject matter experts for accuracy and technicality. Lundgren and McMakin also provide excellent frameworks for audience analysis based upon the functional line of communication being developed; care consensus or crisis communications demonstrate that each require a different type and depth of analysis.

The final critical component is creating messages that overcome influence factors that may or may not be related to the purpose and objectives of the communication event and the audience profile. They take into account the major influences of trust and credibility that have been discussed in previous chapters, which will be referenced repeatedly throughout this text. The latter concept reminds communicators that messages must be crafted to influence an audience's emotional level sometimes as much as they exist to increase knowledge and understanding of the technical aspects of a situation. Of significant help in this task is Covello's risk perception model, which identifies a method for assessing the audience's emotional status related to the risk event at hand.

The difficulties technical experts frequently have in addressing a subject from a nontechnical aspect is related to the above numerous sources on the topic of audience profiling (U.S. EPA 2005). In the EPA handbook that provides guidance for the development of significant community involvement in the Superfund program, the authors chide agency staff:

> Individuals are often much more concerned with nontechnical issues, such as fairness and control, than with the technical details of risk assessment. The risk communicator needs to address both technical risk assessment and nontechnical concerns. Agency representatives have a tendency to focus on the technical issues, often to the exclusion of the public. When this occurs, the Agency representative is not communicating with the public, especially since the public often views risk differently than to the technical experts (U.S. EPA 2005, p. 11).

On the other hand, Morgan cautions communicators not to assume that the audience is unable to understand the technical aspects of risk, even if it does not seem as interested in hearing about them. Too often he argues, some skeptical risk communicators "assert that people are technically illiterate and ruled by emotion rather than by substance—hence education is hopeless" (Morgan et al. 2002, p. 7). Morgan goes on to suggest that the very emotionality of a risk event and its high stakes nature leads an audience to desire to know more and that technical risk communicators often blame the audience when the message is not understood:

> Such emotions need not mean that risk communication is hopeless nor that people are incapable of making reasoned decisions about risks. Indeed, emotion can provide motivation for acquiring competence—even if it makes people more critical consumers of risk communications. Although citizens may begin their learning process with rela-

tively little technical understanding, we believe that most can understand the basic issues needed to make informed decisions about many technically based risks—given time, effort, and careful explanation. Unfortunately when the message is not understood the recipients, rather than the message, often get blamed for the communication failure" (Morgan 2002, p. 8).

It is important to emphasize the comment in the above quote regarding the audience's critical view of the messages they receive. As will be discussed below, analyzing an audience provides information on its members' basic characteristics, which leads to messages that are more effective at achieving the stated purpose and objectives. It does not, however, ensure that the audience members will accept the message, even if they clearly understand the technical content. Risk communicators would do well to remember that the communications events are designed to view the audience as a partner in the effort. Therefore, the audience should have the ability to question or judge the message or disagree. Messages are not simply designed as instructions so that the audience knows how to act and behave. They are designed to be informative to the point where the audience can make assessments of their own, thus leading to decisions about their own actions and behavior, which may be different from what the risk communicator had in mind.

MESSAGE CRAFTING—DETERMINING PURPOSE AND OBJECTIVES

Morgan tackles the idea of the need for goals for risk and crisis messages early on in his discussion of the mental models approach to message crafting (Morgan 2002). One of his key points is that much risk and crisis communication is often done *ad hoc*, with limited planning and forethought before the message is delivered. He raises the flag that this type of message development is risky at best, as the impact on the lives of the audience members can often be significant. It also leads to message delivery that is more likely to fail, leaving the communicator confused and frustrated. Morgan believes that message development rarely involves any audience needs analysis, is rarely analyzed before delivery, and just as rarely is evaluated after the fact to determine whether or not the messages have succeeded. He argues that it is not a difficult process to undertake, and indeed much of this chapter is devoted to the process of message development from beginning to end.

The EPA's Superfund program was one of the earliest attempts by government to develop meaningful partnerships with publics who were affected by the identification, remediation, and monitoring of abandoned hazardous waste sites in their communities. Many of the documents used by the EPA in this program provide excellent reminders of the importance of public involvement in risk communications efforts. Risk communications, as noted in Chapter 2, are not simply opportunities to speak *to* an audience, but to engage the audience in meaningful dialogue and consensus building. In its basic handbook for Superfund community efforts, the EPA notes: "The goal of risk communication is to promote public involvement that is informed, reasonable, thoughtful, solution-oriented, and *collaborative*" (U.S. EPA 2005, p. 10; author's emphasis).

In the same document, the EPA also states: "The goal of the risk communication strategy is to increase the understanding and *involvement* of interested parties in the process rather than reach unanimity" (U.S. EPA 2005, p. 9; author's emphasis).

Finally, the EPA mentions the significance of the establishment of trust and credibility prior to the message delivery (U.S. EPA 2005). This vital component has also been echoed in many of the previously cited and discussed writings of Peters and Covello; Sandman; and Lundgren and McMakin (Peters *et al.* 1997; Sandman 2003 and 2007; Lundgren and McMakin 2004).

Developing message goals and objectives can be as simple as developing a set of key questions to ask as communicators. Rutgers's University Center for Environmental Communication offers the following list (Chess and Hance, n.d.):

1. Why are we communicating?
2. Who are our target audiences?
3. What do our audiences want to know?
4. What do we want to get across?
5. How will we communicate?
6. How will we listen?
7. How will we respond?
8. Who will carry out the plans? When?
9. What problems or barriers have we planned for?
10. Have we succeeded?

While these questions begin to delve into aspects of actual message delivery, their value is in encouraging risk communicators to think through the entire purpose of the message event, not just the content of the message, *before* the message is crafted and delivered. Understanding delivery mechanisms and how to respond to the audience after the message is delivered, as well as the "how" and "when" of the delivery help create an overall message purpose that can then be used to detail objectives and steps to met the goal.

The goals of messages are often fairly simple to identify, if the time is taken to do so. One of the key problems in setting goals is the lack of specificity or realism in a goal that may be identified without fully flushing out some of the parameters. One of the simplest tools available to help develop goals and objectives in nearly every setting is the use of the acronym SMART—Specific, Measurable, Actionable, Realistic, Time—when writing out a goal (Lopper 2006). This technique has been effectively used in a multitude of goal setting activities; risk and crisis messages are no different.

  Specific. The goal must provide detailed information about what the current situation is and what it will be; in other words, how will the message change the situation? Helping the public prepare for a pandemic flu outbreak is not specific. Helping the public to understand the difference between the characteristics of seasonal flu and pandemic flu is much better.

  Measurable. There must be something in the goal that describes a factor that can be measured so that the evaluation of the success of the message can be demonstrated. Getting the public to practice effective social distancing techniques during an outbreak of pandemic flu by reducing the number of group staff meetings by 50 percent is clearly measurable.

Achievable. Goals must be simple enough to be able to be accomplished but also challenging enough that they require some effort to make them successful. A goal that asks for a business to help its staff understand that the technique of coughing into one's elbow to reduce the spread of germs is achievable but also quite simple (and also not measurable). A goal that states a business will place posters that show the elbow coughing technique in all common employee areas such as break rooms and time clocks and will hold 10-minute safety contact meetings for all staff during a specified week to reinforce the technique and its importance is not only achievable but specific and measurable.

Realistic. Many goals that can be identified are excellent ones to achieve, but the end point is so unreasonable that the audience will not try at all or won't try very hard. A goal that asks for businesses to have all of their staff telecommute during a pandemic flu outbreak is not achievable, and many businesses will reject the idea out of hand if presented in that manner. A goal that asks for a business that does not have any employees currently telecommuting to identify at least 25 percent of its staff who can telecommute is. In fact, some businesses might even find more than 25 percent once they begin an honest effort based upon a goal that they see as realistic.

Time. Goals should always set a time frame for completion and those time frames should also be reasonable; otherwise the effort is compromised from the start. A goal of developing and delivering pandemic flu risk messages about social distancing and effective hygiene through classroom-based training sessions to a workforce of 5,000 spread out over five locations in the United States within five business days is not likely to be accomplished.

An additional point about goal setting addresses the need to first develop overall message goals and then identify the concrete objectives that will need to occur in order to achieve the goal. The subsequent planning process that follows the setting of objectives is as important as finding a SMART goal. In the latter example, the goal, once a reasonable time frame is established, will need to be broken up into numerous specific objectives. Those might include convening a team of training curriculum developers from within multiple business units; meeting to determine the messages to be delivered; and identifying the time frames for delivery, including the locations and training of the instructors.

MESSAGE DEVELOPMENT CONSTRAINTS

One last concept to be discussed about message goal setting is the need to understand the constraints of message development. Audience constraint factors are discussed in greater detail below, while the constraints on the communicator due to their own bias about audiences are referenced above.

Recognizing other factors posed by the organization should occur at the earliest possible stage of the process. They should be one of the first areas identified since the ability to deliver a successful message can be completely negated if the organization does not permit or support it, and all the time spent in goal setting, audience profiling, and message crafting will be for naught. Organizational constraints include the following (Lundgren and McMakin 2004):

Inadequate resources. Funding, staff, space, and equipment are basic needs for any risk communicator, but they may not be available or permitted to be used by those who authorize such expenditures. Many, though not all, organizations see the value of resource allocation for risk assessment activities but are not quite convinced that message crafting is any more complicated that deciding what to say and saying it. As this chapter suggests, such is clearly not the case.

Management apathy or hostility. Organizational decision makers may not have a sound understanding of the principles of risk and crisis communications, and thus see no benefit to supporting the effort. As noted in Chapter 2, the principles of risk and crisis communication are less than 25 years old; many senior managers may not have reached an understanding of the practice or its value. Furthermore, apathy may be part of the lack of understanding or may be a totally separate factor related to management's focus on other activities that they view as more vital to the mission of the organization. Even more problematic for the risk and crisis communicator, management's view of risk communication efforts may be negative or even hostile, a significant barrier to overcome before any activities related to audience profiling or message crafting can be undertaken.

Internal organizational processes and procedures. Organizations can sometime lack the ability to move projects and activities through the approval process in a timely manner because the organization is heavy on the management side or simply slow to respond to any new initiative. And some organizations have complicated processes and procedures that require an amount of time that may not always be available when a risk or crisis is looming and messages need to be developed and delivered in a timely manner.

Corporate protection (legal and otherwise). Risk and crisis communications by their very nature require disclosure of information that may not always paint the organization in the best light or may be seen by legal counsel as exposing the organization to liability concerns. Finding a balance between protecting the organization's interests and informing audiences can be a tricky maneuver. In the current litigious climate, erring on the side of caution is more often the chosen method.

PROFILING AUDIENCES—WHO ARE THEY AND WHAT DO THEY WANT?

Once the goals and objectives of the message events are established and agreed to by the organization's management and its communicators, audience profiling is the essential next step. Understanding basic demographic factors such as age and education is part of this process but so is determining the constraints of the message posed by the audience. Communicators need to take the time to discover the audience members' attitudes, beliefs, and emotional state, and give them sufficient consideration in the process. Some questions that should always be asked at the beginning of the profiling process include:

1. What does the target audience want and need to hear?
2. In what format does it want and need to hear it?
3. At what stage of the event should the information be delivered and how often?
4. Is there a person who is the likely messenger to a targeted audience segment?

In complex risk communication processes and crisis events that are comprehensive in their scope and affected audiences, there need to be numerous variations on the key messages to the different audience segments. This practice does not affect the success of the message effort as long as the messages are integrated and contain the same key concepts. (Subsequent chapters will discuss the use of tools such as influence diagrams and message maps and will delve deeper into creating messages based upon key concepts.)

Three possible frameworks appear below as a means of beginning the profiling process. The first provides a method for a generalized grouping of audiences; the second lists some general questions to be asked about the audiences, and the third defines audience profiling on three distinct levels, depending upon the type of message and risk or crisis involved. In reviewing these frameworks, it is helpful to once again point out that message crafting involves a number of key steps, all of which are important and sequential. Profiling the audiences should not be undertaken until the goals and objectives of the message are fully explored and developed.

As an overarching means of segmenting audiences, the categories of publics created by Fearn-Banks (2007) is helpful because it creates a framework for identifying both internal and external publics and recognizes that different messages that must be crafted for each group. She lists four different categories of publics:

1. **Enabling publics.** Publics who run organizations. They have the power and authority to make decisions and include shareholders of publicly traded companies, boards of directors, investors, and top level executives.
2. **Functional publics.** Publics who are responsible for making the organization operate. They include the workforce, suppliers, vendors, contractors, and those who purchase or use the organization's product.
3. **Normative publics.** Publics who share values with the organization and may be partners in the process of risk and crisis communications. They include trade associations, professional associations, and sometimes even competitors.
4. **Diffused publics.** Publics with a vital but indirect link to the organization. The most common examples of this group include the media, community groups, as well as persons who live, work, or play near the organization.

Documents created by the EPA to assist in assessing and communicating risk to key audiences also provide general frameworks that are helpful in the first steps of audience profiling. The EPA offers a set of initial questions to be asked, which include (U.S. EPA 2007):

- What is their current level of knowledge about the risk?
- What do you want them to know about the risk? What actions would you like them to take regarding the risk?
- What information is likely to be of greatest interest to the audience? What information will they probably want to know once they develop some awareness of the risk?
- How much time are they likely to give to receiving and assimilating the information?
- How does this group generally receive information?
- In what professional, recreational, and domestic activities does this group typically engage that might provide avenues for distributing outreach products?
- Are there any organizations or centers that represent or serve the audience and might be avenues for disseminating your outreach products?

Many of the questions above focus more directly on an audience that is composed of external publics, rather than internal. With limited reconfiguration, however, the questions can be modified to assist in profiling an internal audience. For example, the question regarding how the group typically receives information would be answered by how the organization informs its workforce. These traditionally include bulletin boards, safety meetings, and newsletters.

The third framework for general audience profiling comes from Lundgren and McMakin (2004) and considers at the level of profiling needed by identifying three levels of analysis. These levels are sequential in that the amount of information required for each level builds upon the previous one. A general description is provided below; for additional details see Table 4.1.

1. **Baseline audience analysis.** The information required for this level of analysis relates to basic audience characteristics and the ability of the audience to comprehend information. Audience characteristics needed include reading ability, reading comprehension, and preferred methods of communication. Some indications of emotional level, particularly of hostility, are helpful as well. In the discussion of functional lines of communication (see Chapter 3), this level of analysis is necessary for crisis communications; indeed, it may be the only level of analysis that time permits and works because the purpose of the communication event is simply to inform and persuade the audience to take a specified action.

2. **Midline audience analysis.** The information required for this level of analysis includes all of the above for baseline analyses as well as socioeconomic information, demographics, age, gender, and range of occupations. In terms of functional lines of communications, care communications require the informa-

TABLE 4.1. Questions to Ask When Completing an Audience Analysis[a]

Baseline Audience Analysis	
Audience Characteristics	Questions to Ask
Reading level	At what level do they read?
Education level	What is the highest level of education completed?
	What is the range?
Group size	How many people are in the audience?
Information sources	Where do they get information (television news, newspapers, radio, family networks, experience, Internet, social media)?
Background in risk subject matter	How much do they understand about risk scientifically?
Experience with the risk	Is the risk new to the audience members or is this something they have been living with for a long time?
Their "hot buttons"	Are there words and concepts that infuriate them?

Midline Audience Analysis	
Audience Characteristics	Questions to Ask
Age	What age range do they fall into?
	What 5-year range has the most people in it?
Gender	Are they mostly male or female?
Preferred social institutions	Where do they go to relax?
	To play?
	To worship?
Jobs/occupations	Where do they work?
	What do they do there?
	Is the risk part of their workplace?
Geographic areas	How near is the risk?
Length/history of involvement	How long have they been involved with the risk?
	Has the involvement been one of passive listening, consensus building, or reactive argument?

Comprehensive Audience Analysis	
Characteristics	Questions to Ask?
Concerns and feelings about risk	What kinds of concerns do they have?
	How do they feel about the risk (angry, frustrated, apathetic)?
Experience with other risks	Have they had good examples you can build on?
	Bad examples to overcome?
Exposure to news media or other coverage	Have they seen comprehensive coverage or tabloid-style journalism?
Goals of organized groups	What are they trying to accomplish?
Effect of the risk on them	How do experts think the risk can affect them?
	How do audience members think it can affect them?
Their control over the risk	Can they mitigate the risk or must they live with it?

[a]Lundgren and McMakin (2004), Table 8.3 (pp. 130–132)

tion provided by this higher level of analysis; the goal is to increase awareness for future purposes, which include taking action and reaching a consensus.

3. **Comprehensive audience analysis.** Building upon the information gathered for baseline and midline analyses, this level addresses more abstract factors such as motivations for receiving information as well as perceptions of risk in terms of vulnerability and control. Consensus and care communications typify this level of analysis; in order to change perceptions and change behavior more substantial information about the audience's thoughts and feelings about the risk is crucial.

A final, but vital component of initial audience profiling involves an assessment of what the audience wants from the communication event. Audiences receive and respond to messages that are crafted with their needs and desires in mind. These messages must also take into consideration the type of information the audience wants and the level of participation the communication event provides it. As was noted in the beginning of this chapter, the type of information risk assessors and communicators want the audience to receive is not always what it wants, or more importantly, are willing to respond to.

PROFILING AUDIENCES—HOW DO THEY PROCESS AND PERCEIVE THE RISK?

Morgan (2002) identifies some key issues to consider by reminding us that an audience is more often concerned with immediate hazards and concerns, while scientific professionals are more often involved in rare and unusual hazards. He also notes the tendency of an audience's alignment of risks to be related more to the hazards it most often hears about rather than the ones that might be more likely to affect it, all of which is further compounded by the influence of contemporary media. For example, media accounts of horrific shark attacks can clear beaches, despite the relatively minor risk of being personally involved in a shark attack, while the more common risk of crashes from distracted driving do not seem to change risky behaviors such as talking on a cell phone while driving or reading directions on a piece of paper. This misalignment with the audience's quantitative perception of risk should give risk communicators reason to carefully consider their delivery of complicated statistical analyses, as these types of communications require a level of audience analysis that provides significant information about education and reading levels, and they require communicators to craft messages that help the audience comprehend the statistical information and use it to evaluate the risk posed to it.

Morgan (2002) also further flushes out an audience's perception of risk by addressing factors beyond sensationalism. These include how well the risk is understood, how equitably the risk is distributed across the population, how well individuals can control the risk, and whether the risk is assumed voluntarily or is imposed on people without their approval. The table from Chapter 3—Covello's 15 Risk Perception Factors—is provided again below (Table 4.2) and illustrates these factors, as well as additional ones that are outlined in Covello's risk perception model.

TABLE 4.2. Covello's 15 Risk Perception Factors[a]

Risk Factor	Applicability
Voluntariness	If the audience members perceive the risk to be voluntary, they are more likely to accept it because they understand their role in experiencing the implications of the risk.
Controllability	If the audience members perceive that they have control over the risk, they are more likely to accept the implications of it.
Familiarity	If the audience members have some previous knowledge of the risk or experience with it, they are more likely to accept the implications of it because of the increased level of knowing what might or might not happen.
Equity	If the audience members perceive the implications and consequences of the risk to be equally shared among audience members, they are more likely to accept the implications of it.
Benefits	If the audience members perceive the ultimate benefits of the risk to be positive, they are more likely to accept the potential negative implications of experiencing it.
Understanding	If the audience members possess a basic understanding of the risk, they are more likely to accept the implications of it. The greater the level of understanding, the higher the acceptance.
Uncertainty	If the audience members perceive the risks have a degree of certainty in various dimensions and in the scientific information available about it, they are more likely to accept the implications of it.
Dread	If the audience members' emotions with regard to a risk are less intense and fearful, the more likely they are to accept the implications of it.
Trust in institutions	If the audience members perceive the institutions more significantly involved in the risk as trustworthy and credible, the more likely they are to accept the implications of it.
Reversibility	If the audience members perceive the risk to have reversible adverse effects, they are more likely to accept the implications of it.
Personal stake	If the audience members perceive the risk to be limited in its personal implications and consequences, the more likely they are to accept the implications of the risk.
Ethical/moral nature	If the audience members perceive the risk to be morally or ethically acceptable, they are more likely to accept the implications of it.
Human vs. natural origin	If the audience members perceive that the origin of the risk is naturally occurring, they are more likely to accept the implications of it.
Catastrophic Potential	If the audience members perceive that the amount of fatalities, injuries, and illnesses from a risk are minimal, they are more likely to accept the implications of it.

[a]Covello et al. 2001

In addition to thinking about how audiences as a whole perceive risk, it is also important to consider the influences of an individual audience member's perception of risk. Two factors predominate: personal experience with the risk and media coverage. Personal experience can occur in the audience member's personal life, work life, or both; however, most people tend to develop stronger emotional reactions to those risks that affect their personal life. The assumption here is that working has inherent risks, regardless of the job type, and, in many situations, the higher the risks the better the compensation for the position. But many audiences will also understand that the individual typically has some level of choice about their chosen occupation. Therefore, the control of the risk rests to some degree with the individual along with the ability to remove the risk from their lives. However, the adage that "a man's home is his castle" is more universally understood and appreciated. Therefore, anything that invades that space and creates risk gives rise to strong emotions and the desire to defend the castle, especially if the risk affects the safety, health, and lives of an audience member's family.

An additional and often related factor in recognizing the influences of risk perception in an audience is the tendency to prioritize a personal list of risks based upon recall of direct and indirect experience with those risks. Health problems like heart disease and stroke are serious and affect millions of families each year, either because an individual suffers from the disease or because a direct family member does. This risk may appear on an individual audience member's list of concerns, but will likely rise in importance given any recent personal experience with either health problems through hospitalizations or even death.

The influence of the media is noted above when considering audience perception of risk, but the influence of media due to the level of coverage of a risk has grown substantially in the past 20 years as media information sources dominate the lives of audiences. Not only are messages coming from new and different sources (primarily digital sources such as the Internet), but the flow of coverage is constant and unrelenting. The substantial number of messages about any given risk can substantially alter audience members' perceptions of its importance and likelihood it will affect their lives. In addition, current media sources may lack the objectivity that was more common in the recent past and may even be patently untrue, yet presented in a believable way or at least perceived to be true by some segments of the audience. Counteracting these reliable sources of information will be addressed in greater detail in Chapter 7; however, doing so can be difficult, at best.

REFERENCES

Chess, C. and B.J. Hance. n.d. "Communicating With the Public: Ten Questions Environmental Managers Should Ask." Center for Environmental Communication, Cook College, Rutgers, The State University of New Jersey, New Brunswick, NJ.

Covello, V., R. Peters, J. Wojtecki, and R. Hyde. 2001. "Risk Communication, the West Nile Virus Epidemic, and Bioterrorism: Responding to the Communication Challenges Posed by the Intentional or Unintentional Release of a Pathogen in an Urban Setting." *Journal of Urban Health* 78(2):382–391.

Fearn-Banks, K. 2007. *Crisis Communications: A Casebook Approach*, 3rd ed. Mahwah, New Jersey: Lawrence Erlbaum Associates.

Lopper, J. 2006. "SMART Goal Setting." Posted online at http://www.suite101.com/content/ smart-goal-setting-a9911. Accessed on February 6, 2010.

Lundgren, R.E. and A.H. McMakin. 2004. *Risk Communication: A Handbook for Communicating Environmental, Safety, and Health Risks*, 3rd ed. Columbus, OH: Battelle Press.

Morgan, M.G., B. Fishhoff, A. Bostrom, and C.J. Atman. 2002. *Risk Communication: A Mental Models Approach*. New York: Cambridge University Press.

Peters, R.G., V.T. Covello, and D.B. McCallum. 1997. "The Determinants of Trust and Credibility in Environmental Risk Communication: An Empirical Study." *Risk Analysis* 17(1):43–54.

Sandman, P. 2003. "Four Kinds of Risk Communication." Posted online at http://www. petersandman.com/col/4kind-1.htm on April 11, 2003. Accessed on January 4, 2010.

Sandman, P. 2007. "Empathy in Risk Communication." Posted online at http://www. petersandman.com/col/empathy.htm on July 29, 2007. Accessed on January 11, 2010.

U.S. Environmental Protection Agency. 2005. "Superfund Community Involvement Handbook." EPA 540-K-05-003.

U.S. Environmental Protection Agency. 2007. "Risk Communication in Action: The Risk Communication Handbook." EPA 625-R-05-003.

5

CRAFTING RISK AND CRISIS MESSAGES— DEVELOPING THE WORDS

This chapter will continue with a review and explanation of several key techniques for message crafting, including influence diagrams and message mapping, and will provide some additional discussions about message delivery. The chapter will close with a review of some of the most common errors made by communicators.

CRAFTING MESSAGES—OVERARCHING PRINCIPLES

It is helpful at this juncture to restate some of the overarching principles regarding risk and crisis messages that have been detailed in previous chapters. These key concepts are summarized by Covello as follows (Covello *et al.* 2001, Covello 2008):

- Providing knowledge needed for informed decision making about risks
- Building or rebuilding trust among stakeholders
- Engaging stakeholders in dialogue aimed at resolving disputes or reaching consensus
- Minimizing conflict among messengers and messages

Risk and Crisis Communications: Methods and Messages, First Edition. Pamela (Ferrante) Walaski.
© 2011 John Wiley & Sons, Inc. Published 2011 by John Wiley & Sons, Inc.

- Good planning on message content
- Delivering messages with skilled practitioners

In addition to the above list, Covello also addresses the need to acknowledge uncertainty, which is also noted by other practitioners (Covello 2008; Sandman 2007; U.S. DHHS 2006), with the reminder to risk and crisis communicators that the complexity of the data, as well as its incompleteness, creates obstacles that must be overcome. Covello suggests the following strategies:

- Acknowledge the uncertainty, rather than hide it
- Explain to the audience about the difficulties of assessing and estimating risk
- Provide clear and simple explanations for how the data was obtained
- Share information promptly with appropriate reservations about uncertainty
- Clarify for the audience what information is certain, nearly certain, not known, may never be known, is likely, is unlikely, is highly improbable, and what is being done to reduce the level of uncertainty
- Correct any errors in data or clarifications as soon as they become known

As noted above, despite the demand from the audience for comprehensive and reliable information about a risk or a developing crisis, the reality faced by communicators is that the information desired by the audience may not always be available or known. Even the best available scientific data is not always complete, and those unfamiliar with statistical analysis often assume causality in situations where the data simply establish a relationship. An excellent example of this concept is the direct correlation of the increase of the consumption of ice cream and the number of sexual assaults. A review of basic data regarding these variables establishes a positive relationship but does not provide causation. Further analysis of the variables reveals that both are related to a confounding variable: weather. The hotter the temperature, the more ice cream is consumed; the hotter the weather, the more likely people are to leave windows open in their homes, making easier access for sexual assaults. While this example provides a simplistic comparison, audience members and even communicators who do not possess a clear understanding of statistics may have a harder time understanding the finer points and may resort to simply reporting data. It has been suggested that an audience demanding 100 percent certainty in the reporting of data is not typically questioning the numbers but the processes that were used to obtain the numbers. Explaining the latter can often reduce or eliminate the audience demand for absolute certainty. This concept is also tightly bound to previous discussions of trust and credibility. Communicators with a reputation for being credible and trustworthy are more likely to have their data received positively, and the answers they give to questions about the data are more likely to be believed.

Along the same lines, Sandman discusses the importance of being able to say, "I don't know" during a communication event, viewing it as a credibility builder rather than detractor (Sandman 2004a).

CONVEYING EMPATHY

All of the possible risks or crises that require communication events to audiences also require message delivery that addresses two key concepts: The first is the ability of the communicator to demonstrate empathy with the audience, and the second is for the communicator to be able to deal competently with the predominant emotion of the audience, typically anger, fear, panic or apathy, and mistrust.

Several authors address the subject of empathy. Coombs use the term "sympathy," although in his description of the concept, he indicates that he means compassion for the "victims" of the event (Coombs 1999). He hastens to add that sympathy does not translate to an acceptance of responsibility for liability on the part of the organization. He also expands on his perception of the two vital concepts of trust and credibility and suggests that a successful demonstration of sympathy by the communicator creates and/ or enhances both concepts.

Covello also writes about the importance of empathy and suggests that audiences "want to know that you care before they care about what you know" (Covello 2008). The ability to demonstrate empathy in the earliest stage of message delivery helps to reduce the stress that is nearly universal in all risk and crisis communication events and allows an audience to improve its ability to hear, understand, and remember information that might be scary or dangerous.

Finally, Sandman has devoted several of his major writings to the importance of demonstrating empathy in risk and crisis communications, including one substantive article on the topic of empathy alone (Sandman 2007). He strongly supports Covello's assertion noted above that it is important to show you care, and until audience members know the communicator cares, they will generally be incapable of hearing, reacting, and responding to a risk or crisis communication message. As noted briefly in Chapter 3, empathy is not always something that is intuitive to risk communicators; in fact, Sandman suggest that most risk communicators' intuition about empathy is incorrect and can often do more harm than good. While it may be easy to craft messages that have the correct content, assuring that they have the correct emotion is not simple at all.

Sandman's illustrative comment that "people are often very proprietary about their pain" (Sandman 2007, p. 1) provides some guidelines for showing empathy. He stresses that empathy is a feeling, not a strategy, and cannot be part of a checklist used to craft messages. It stands to reason that risk communicators ought to be sincere in their concern for the audience when discussing risks to it and that even if the communicator does not fully understand an audience's feelings about an event that threatens it, the communicator can at least be honest about trying to understand.

Communicators will likely experience failure if they also attempt to convince audience members that they should not be feeling a specific emotion; in fact, this strategy often backfires and creates a situation where an audience's emotional reaction strengthens in response and becomes even more firmly entrenched. If the audience's emotions are not in line with the reality of the situation, a communicator may have inadvertently created a more difficult situation, setting up any future communications efforts to be substantially harder than they need to be.

The essence of empathy is what Sandman terms a "sort of acknowledgement." It is a key component is assessing how audiences feel about the situation and involves acknowledging the audience's strongest feelings, often of anger, betrayal, and anxiety, and is critically tied to the ability of the communicator to demonstrate caring and concern. It also involves recognizing how the audience feels about the communicator in both positive and negative ways. It involves who the audience thinks the communicator is, both in the primary role as a communicator and as a fellow human being. Finally, it involves allowing for how the audience members imagine the communicator feels about them.

According to Sandman, a core task in risk and crisis communication events is learning how to disagree empathetically. He suggests that communicators not attempt to spend as much time trying to show they care and agree with the audience's perception even when they don't, but to look for areas of agreement upon which to build common understanding and acceptance of each other's feelings. This strategy is more open and allows for honest disagreements about the risk or crisis situation. Disagreeing empathetically, according to Sandman, is the crucial combination by a communicator of demonstrating sincere empathy about the audience's emotions and feelings about an event, along with appropriate levels of candor about their own feelings and emotions about the event when they don't match up. He says:

> Even though candid statements about how things seem to you are obviously distinct from empathic statements about how things might seem to your stakeholders, the essence of risk communication is to find ways to combine the two. Your goal as a communicator is very often to change your stakeholders' minds about something. So empathy alone won't accomplish your goal; at some point you're going to need to tell them what you think and why. But candor alone probably won't accomplish your goal either. If you're not attuned to their perspectives, they're profoundly unlikely to accept yours" (Sandman 2007, p. 10).

In Chapter 3, Sandman's types of risk communication were explained (outrage management, precaution advocacy, and crisis communications) (see Fig. 5.1). Each one of these types requires a slightly different understanding of how to show empathy:

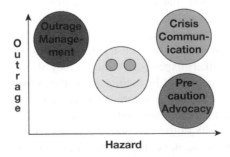

Fig. 5.1. Three Types of Risk Communications According to Peter Sandman

- **Outrage management (low hazard—high outrage).** When the hazard is low but audience members' emotional level concerning it and their assessment of the level of danger to themselves is high, the use of concrete data and statistical information to change their opinions of the risk are highly unlikely to succeed. This task is made even more difficult if the communicator is a member of the organization deemed by the audience to be responsible for creating the risk in the first place.

- **Precaution advocacy (high hazard—low outrage).** In this type of risk communication, the audience's denial or apathy is the predominant factor and increasing the outrage to match the hazard level can be difficult. When the audience doesn't appear to care or isn't paying attention to the message, expressing empathy and concern will likely go unnoticed. But that does not negate the importance of the communicator making the attempt to understand the emotional state of the audience. Doing so will help the communicator determine how best to increase the audience members' emotional level with regard to what appears to be a true hazard to them, based on the available data. The tricky part of this scenario is that risk communicators may have a difficult time achieving that understanding because their professional lives have been deeply involved in learning about and understanding the hazard data. By default they tend to believe in it and may sincerely wonder how someone else would not. A more in-depth discussion of how an audience's apathy may be masking depression or fear is presented below.

- **Crisis communications (high hazard—high outrage).** As was noted in the quote from Mayor Rudy Giuliani shortly following the attack on the World Trade Center in 2001, communicators can increase their chance at achieving communication objectives by acknowledging their own fears and worries as well. Attempting to hide them or pretending they do not exist can serve to increase the audience's level of fear or add the element of anger to the emotional state of the audience. Even more, trying to over-reassure the audience that the situation is not nearly so bad, when they can clearly see the risk, is dangerous and ineffective. Sandman notes, "Crisis managers who imagine that showing empathy means over-reassuring people, 'emphasizing the positive,' or 'calming them down' are way off the mark" (Sandman 2007).

AUDIENCE EMOTIONS—ANGER

The strongest emotions generally exhibited by an audience, and those that tend to derail communication events most often include anger; fear, panic, or apathy (all relative emotions and addressed together); and mistrust. Being aware of which of these emotions are most likely to be in place prior to a communication event helps craft messages that contain the correct amount of technical information to help the audience understand the hazard, as well as set the correct tone to demonstrate to the audience that its emotional state is both understood and validated.

However, communicators would do well to take stock of their own level of anger before attempting to mitigate that of the audience's. An angry communicator is often unable to mask the emotion for any extended period of time and may be very vulnerable to harmful and "knee-jerk" responses that include sarcasm and passive-aggressiveness. It is often true that if the audience is angry at the communicator (or the organization he/she represents), the communicator is also angry at the audience (Sandman 1995b).

Many communicators would like to believe that they are able to simultaneously present factual information dispassionately for situations that are truly risky, while being assailed by the audience and refraining from displaying any emotion. Part of this mistaken assumption may be related to the self-esteem level of the communicator and part may be a mistaken belief on the part of the communicator that being professional also means being devoid of emotions and that the audience expects and desires this type of detachment from communicators. As was noted above, this myth contributes to the delivery of messages that are unable to demonstrate empathy for the audience's situation, which the commentator may also share. In addition, risk communicators may lack training and experience in diffusing highly charged emotional situations; this may inadvertently contribute to the increase of emotion by trying to downplay the risk, also as noted above.

Communicators are sometimes trapped by mistakenly assessing an audience's emotional reaction as one of greed rather than anger. Sandman illustrates this concept with the use of litigation by various audiences and suggests that, for the most part, an audience involved in litigation as a plaintiff is more often angry and desires revenge than it is greedy (it may be the lawyers who are "greedy", however). The difference may be subtle but can be better understood by considering what the audience views as a "win" and as a "loss." Sandman's perspective is that greed is typically displayed as wanting to "win money" from the organization, and anger shows up as the audience wanting the organization to "lose money" (or prestige or market share or some other tangible loss directly to the organization) (Sandman 1995a).

Critical to understanding an audience's anger is being able to put it into some type of framework that allows assessment of the various factors that are involved. As noted in Chapter 3, Covello's risk perception model provides a list of 15 different factors that help identify the level of risk perceived by an audience. These factors also provide a framework for assessing the audience's anger level as well (see Table 5.1).

In general, those factors that tend to increase an audience's level of anger at both the situation and the organization represented by the communicator include the level of voluntariness and control over the situation. For example, if the ultimate control over the audience's exposure and danger from the hazard is under the control of the organization or predominantly so, the level of anger can be expected to increase dramatically. To rectify this situation, the organization needs to effectively communicate that the audience's perception of the amount of control the organization had over the hazard's occurrence in the first place was off the mark or at least that the organization is doing all it can to reduce the risk to the hazard.

Additional factors that strongly influence audience anger levels include whether or not other members of the audience are equally at risk, particularly if the subgroup the

TABLE 5.1. Using Covello's Risk Perception Factors to Evaluate Audience Anger Potential

Risk Factor[a]	Applicability[a]	Affect on Audience Anger Level[b]
Voluntariness	If the audience members perceive the risk to be voluntary, they are more likely to accept it because they understand their role in a experiencing the implications of the risk.	High
Controllability	If the audience members perceive that they have control over the risk, they are more likely to accept the implications of it.	High
Familiarity	If the audience members have some previous knowledge of the risk or experience with it, they are more likely to accept the implications of it because of the increased level of knowing what might or might not happen.	Low
Equity	If the audience members perceive the implications and consequences of the risk to be equally shared among audience members, they are more likely to accept the implications of it.	High
Benefits	If the audience members perceive the ultimate benefits of the risk to be positive, they are more likely to accept the potential negative implications of experiencing it.	Low
Understanding	If the audience members possess a basic understanding of the risk, they are more likely to accept the implications of it. The greater the level of understanding, the higher the acceptance.	Low
Uncertainty	If the audience members perceive the risks have a degree of certainty in various dimensions and in the scientific information available about it, they are more likely to accept the implications of it.	Moderate
Dread	If the audience members' emotions with regard to a risk are less intense and fearful, they more likely they are to accept the implications of it.	Moderate
Trust in institutions	If the audience members perceive the institutions more significantly involved in the risk as trustworthy and credible, the more likely they are to accept the implications of it.	High
Reversibility	If the audience members perceive the risk to have reversible adverse effects, they are more likely to accept the implications of it.	Low
Personal stake	If the audience members perceive the risk to be limited in its personal implications and consequences, the more likely they are to accept the implications of the risk.	Low

(*Continued*)

TABLE 5.1. (Continued)

Risk Factor[a]	Applicability[a]	Affect on Audience Anger Level[b]
Ethical/moral nature	If the audience members perceive the risk to be morally or ethically acceptable, they are more likely to accept the implications of it.	High
Human vs. natural origin	If the audience members perceive that the origin of the risk is naturally occurring, they are more likely to accept the implications of it.	Low
Catastrophic potential	If the audience members perceive that the amount of fatalities, injuries, and illnesses from a risk are minimal, they are more likely to accept the implications of it.	Low

[a]Covello et al. 2001
[b]Author generated

audience member belongs to is at greater risk and the perceived morality of the risk. The latter is, of course, more difficult to ascertain due to the wide range of moral positions taken on any one issue.

Lundgren adds additional depth to the assessment of an audience's anger level by noting that organizational credibility also impacts the anger level, a concept discussed in Chapter 4 as well as in the following section on mistrust. Lundgren also notes that audiences who feel they are being placated by the communicator are more prone to anger, once again demonstrating the importance of empathy. Finally, she notes that audiences who feel as though their concerns are being ignored, who feel that they are being asked to make significant life changes, and who are having difficulties understanding the risk or the data being presented are more likely to be angrier (Lundgren and McMakin 2004).

Effectively dealing with an audience's anger involves addressing multiple concerns and trigger points, usually at the same time. Critical for the individual communicator is the ability to not take personally the audience's hostility and expressions of anger, even if the audience is angry at the organization the communicator represents. Strategies advanced by the U.S. Department of Health and Human Services include (U.S. DHHS 2006):

1. Acknowledge the existence of hostility. The worst thing you can do is pretend it's not there.
2. Practice self-management. Send the message that you are in control.
3. Control your apprehension. Anxiety undercuts confidence, concentration, and momentum.
4. Be prepared. Practice your presentation and anticipated questions and answers.
5. Listen. Recognize people's frustration and communicate empathy and caring.
6. Assume a listening posture. Use eye contact.

7. Answer questions thoughtfully. Turn negatives into positive and bridge back to your messages.

AUDIENCE EMOTIONS—MISTRUST

While anger is perhaps the most common intensive emotion displayed by an audience during risk and crisis communication events, mistrust is also one that surfaces quite readily. Although both emotions are related and often occur together, they can also exist separately. As was noted in Chapter 3, representatives of large corporations and government are more often the recipient of mistrust (Peters *et al.* 1997). Communication efforts that are successful in defying the stereotypes widely held by many audiences are more likely to reduce the level of mistrust. For government representatives, the perception seems to be that pure political ambitions and the desire to remain in power guide most decision making. Politicians who vote against their party despite substantial pressure would send a message to a potential audience that would help reduce this stereotype. In addition, activities by a governmental agency that move quickly to resolve an issue without a lot of bureaucratic difficulties or do not appear to be guided by the political party in power would also help. For corporations, admissions of responsibility for problems caused by action or inaction would likely improve the trust and credibility level as would those efforts designed to improve the community, such as beatification projects or sponsorship of community events.

In 1988 Covello and Allen provided a summary of five strategies for building trust and credibility. (Covello *et al.* 1988) These are summarized below:

1. Accepting and involving the public as a partner helps to allay fears, dispel misinformation, and build consensus on the best methods to mitigate risks.
2. Appreciating the public's specific concerns to convey a sensitivity to normal human fears and worries. Empathy and shared emotions provide needed answers and information and are respectful.
3. Being honest and open helps the public understand the risk and learn how to protect themselves. Misleading information later discovered makes regaining trust and credibility nearly impossible.
4. Working with other credible sources reduces confusion and mistrust and increases the chance that the audience will both accept the information and take appropriate actions.
5. Meeting the needs of the media not only helps them to inform the public but will be done with or without the communicator's assistance. Help the media ensure the information they provide is accurate and enlightening.

AUDIENCE EMOTIONS—FEAR, PANIC, AND APATHY

Lundgren and Sandman both deal with the emotions of fear, panic, and apathy (Lundgren and McMakin 2004; Sandman 2003 and 2004b). These related emotions are often aroused when factors such as dread and lack of control are in place. Other factors of importance include risks that are exotic and unknown or unknowable and, to a lesser degree, those for which the audience's memory of a past similar risk is recent or strong.

Both Lundgren and Sandman agree that true panic is a rare occurrence in an audience. The activities of a panicked audience include behavior that is ultimately damaging to themselves and others; visions of large numbers of people running through the streets destroying property and engaging in acts of violence come to mind. Credible reports of these types of situations are unusual, but memorable, which may be what causes some risk communicators to be more concerned about panic than is warranted. Panic also produces such strong emotions that a person is often unable to respond and freezes. This type of behavior is usually temporary, however, and is in part why acronyms and phrases to help "jar" a person into beginning to take action to break the wave of panic are used for emergency response training. (e.g., using the "PASS" acronym—Pull, Aim, Squeeze, Sweep—to train on the use of a fire extinguisher or teaching children whose clothing may have caught fire to "stop, drop, and roll" rather than run are examples.) Panicky behaviors and thoughts are more often temporary, but when they occur they can effectively block any risk or crisis message from even being heard, let alone be acted upon.

Sandman also raises another variation of a risk communicator's concern about audience panic: panicking an audience more than you want to with the risk communication message. In other words, he is talking about messages designed to move the outrage level that succeed more than expected, thereby moving the outrage level of the audience into panic. While Sandman agrees this possibility exists, he reminds again that it is rare. Using it as a reason for not providing honest and meaningful information to an audience is not only foolish but unethical. Determining whether to withhold information from an audience, how much, and when is nearly impossible to be determined within any specific moral context. Being less than honest with an audience for any reason also goes to the issue of a risk communicator's credibility, an aspect that has been discussed all along in this text. When an audience discovers that the information they have been given is less than correct, for any reason, a risk communicator's credibility, and that of the organization he or she represents, will be difficult to recover. For that matter, audiences are more likely to take the risk communicators' opinions about the risk and actions less seriously. Sandman argues that the paradox of discussing scary information about worst-case scenarios actually works in the opposite direction. He suggests the audience has probably already thought about the worst that can happen; acknowledging it increases the communicator's credibility and enhances the possibility that the audience will take the prescribed action. The following summarizes this point:

> People simply don't tend to overact to honest information about high-magnitude low-probability risks. There may be a brief period of "over-reaction" as they get used to the idea. But a public that has adjusted to the possible worst-case scenario is more stable than a public that doesn't know about it yet—and more calm than a public whose worst case worries are solitary, secret, and uniformed (Sandman 2004b, p. 6).

More often what communicators refer to as panic is a variation of this emotion—denial, which can be a true obstacle for risk and crisis communicators. Denial refers to the shut down of coping behaviors and the suspension of active listening. Audience members in true denial may appear to be listening and may even agree to behave in a certain way but are not actually capable of either. More often than not, an audience in this emotional state will be quiet and, as Lundgren suggests, "deceptively calm" (Lundgren and McMakin 2004, p. 63). Risk communicators who make assumptions in these situations often are surprised and disappointed at later points in the crisis situation. Lundgren suggests an intensive audience profile as the best means to better understand the audience's true emotional state in order to craft messages designed to achieve understanding of the message content and behavior change.

Sandman suggests that denial occurs when audiences simply cannot bear their emotions because they are too frightening and even, on occasion, hurtful and guilt producing. People who ignore obvious signs of health problems but refuse to seek medical attention are classic examples of this type of behavior. Mayor Rudy Giuliani's response to the reporter's question about the number of 9/11 causalities mentioned in Chapter 3 is an excellent example of how to help audiences move out of denial by allowing them to incrementally bear their difficult emotions. Being able to slowly develop a sense of shared fear among an audience is an effective strategy for this type of situation. The phrase "action binds anxiety" is another strategy to employ. When situations are truly frightening or difficult to bear, keeping audience members from sliding into denial by getting them to do something, even something seemingly trivial, is an effective first step. Telling the audience members what warning signals they can be on the alert for and what they can do to protect themselves releases members from their passivity and encourages them to take responsibility for their own safety. This type of strategy is behind most home emergency preparedness activities that may require an audience to temporarily rely wholly on themselves for their day-to-day needs until the government can ensure its ability to provide essential services and return to tending to public safety (U.S. FEMA 2004). Even more effective is developing shared strategies that are agreed upon by communicators and audience members alike (Sandman 2003, 2004b, 2007). As has been alluded to previously, true risk communication efforts invoke this type of partnership and consensus building between the organization and the affected audiences.

Apathy can occur when an audience does not understand the true nature of the risk. As has been previously suggested, this can result from risk assessments that focus on statistical analyses that are quite complicated, and thus difficult to explain in a manner the lay public will be able to understand. In these situations, the risk communicator must regroup and find alternative methods to communicate the risk that are understandable, often a difficult task for professionals who may be so entrenched in their field of study that they are unable to discuss it in simple language and terms. They are genuinely confused and sometimes frustrated when they believe they are communicating a concept clearly but which may be only understandable to those with their background and experience. Apathy can also result from risks that are so far removed from the typical daily lives of audience members that the risks appear to be unworthy of the audience's concern, much less action. In these exotic and unfamiliar situations, communication strategies that bridge the gap between the audience's understanding of the unfamiliar to the familiar will be helpful.

TABLE 5.2. Factors That Can Affect an Audience's Emotional Level

Safe[a]	Risky[a]	Outrage Appears As . . . [b]
Voluntary	Coerced	Anger/Mistrust
Natural	Industrial	Anger/Mistrust
Familiar	Exotic	Fear
Not memorable	Memorable	Denial/Apathy
Not dreaded	Dreaded	Denial/Apathy
Chronic	Catastrophic	Fear
Knowable	Unknowable	Fear
Individually controlled	Controlled by others	Anger/Mistrust
Fair	Unfair	Anger/Mistrust
Morally irrelevant	Morally relevant	Denial/Apathy
Trustworthy sources	Untrustworthy sources	Anger/Mistrust
Responsive process	Unresponsive process	Anger/Mistrust

[a]Sandman 1991

[b]Author generated

Finally, apathy can also occur when an audience hears the risk assessment and understands it, but simply disagrees. While many risk assessments are quantifiable, some are less so, and individual audience members can come to different conclusions about the data, even when they understand it. In these situations, simply repeating the data won't necessarily change the audience's understanding and perception of it, but finding commonalties in the risk perception of the communicator and audience may be effective. This is the type of ongoing dialogue and consensus-building activities most often promulgated by the EPA in its activities with communities in much of its work with the Superfund program, true risk communications as noted above. Other strategies are similar to those discussed in Sandman's concept of precaution advocacy (high hazard-low outrage), discussed above and in Chapter 3.

As can be seen from Table 5.2, combining Sandman's 12 principle components of outrage (displayed in an audience emotions such as anger, fear/panic, denial, apathy, and mistrust) into one chart helps summarize the various factors that play significant roles in an audience's emotional level, thereby providing critical keys to crafting messages that both address those emotions, work with them, and move the risk communication effort forward.

MESSAGE-CRAFTING TECHNIQUES

Developing the specific words of the messages that are delivered in a communication event can only take place after the goals and objectives have been determined, the constraints of the various parties have been identified and addressed, and the audience members have been profiled to assess both their knowledge level and their emotional state. Skipping any of these sequential steps is likely to lead to varying degrees of message failure. The remaining sections of this chapter will address two message-crafting techniques developed by Morgan (influence diagrams) and Covello (message maps). Before that, a few summary points of message content will be helpful: These "rules" were developed by Covello and Allen in 1988 for use in Superfund community activities but are applicable to nearly all risk and crisis communication efforts (Covello *et al.* 1988). (Note: these seven rules are similar to those referenced above with regard to building trust and credibility but add several others to be more comprehensive to all types of situations.)

1. **Accept and involve the receiver of risk information as a legitimate partner.** People have the right to participate in decisions that affect their lives.
2. **Plan and tailor risk communication strategies.** Different goals, audiences, and communication channels require different risk communication strategies.
3. **Listen to your audience.** People are usually more concerned about credibility, competence, and empathy than they are about risk levels, statistics, and details.
4. **Be honesty, frank, and open.** Trust and credibility are difficult to obtain; once lost they are almost impossible to regain.
5. **Coordinate and collaborate with other credible sources.** Conflicts among organizations make communication with the public more difficult.
6. **Meet the needs of the media.** The media play a major role in transmitting risk information. It is critical to know what messages the media deliver and how to deliver effective risk messages through the media.
7. **Speak clearly and with compassion.** Technical language and jargon are major barriers to effective risk communication. Never let your efforts prevent acknowledgement of the tragedy of an illness, injury, or death.

INFLUENCE DIAGRAMS—THE MENTAL MODELS APPROACH

Morgan and his colleagues at Carnegie Mellon University in Pittsburgh, Pennsylvania, have developed a model approach to the development of risk communication messages, particularly those in print format. Their approach focuses on a five-step process and has been used to develop national public health documents on radon, HIV/AIDS, climate change, and other environmental issues that affect the majority of the U.S. population. Their approach is time consuming and expensive, and although it may be beyond the scope of lesser types of risk communication efforts, it provides a helpful framework for message crafting (Morgan *et al.* 2002).

The mental models approach surmises that an audience often possesses gaps in the knowledge base regarding many aspects of a particular risk as it relates to the audience. Morgan postulates that risk and, to some extent, crisis communication messages, are more effective if the time is taken to identify the gaps, based upon a solid foundation of knowledge from experts in the specific topic area. Through the use of an expert model, called an influence diagram, all aspects of a risk can be identified and can be interconnected by various linkages. These models are created with the help of several experts in the topic area and are reviewed by still more to ensure completeness and consensus on the components of the model.

The audience's level of knowledge is then determined using the influence diagram to conduct open-ended interviews with a small cross-section of the intended audience. This step helps to identify correct and incorrect or incomplete beliefs among the audience about significant aspects of the risk. From these initial open-ended interviews, revisions are made to the influence diagrams, again using experts in the topic area. Formal structured interviews are developed and are administered to larger groups. The audience's beliefs and knowledge base become clear, and this leads to the construction of the risk communication, which is then delivered and evaluated.

Two examples of influence diagrams appear below as Figs. 5.2 and 5.3 (Morgan *et al.* 2002).

In an influence diagram, arrows or "influences" are used to show connections between the individual nodes. The simplest of influence diagrams use only ovals, which represent uncertain circumstances, and rectangles, which represent choices made by the decision maker. Arrows denote influence from the initial node to the ending node. Figure 5.2 above shows a representation of the risk of falling down the stairs by a home's resident. The essential circumstances for the fall appear in a line down the center to which additional factors, such as children leaving toys on the stairs, have been added to stretch the interconnected chain out further. Decisions about how to reduce the risk (disciplining the children) appear around the circumstances nodes as appropriate and sometimes influence more than one circumstance, as is seen with the decision to remodel the house.

Figure 5.3 expands upon the simple influence diagram above and relates more to the types of risk analysis communicators should have experience with prior to crafting their messages; these are necessarily more complicated and interconnected. In this example, four levels of circumstances are all interconnected and form a chain of events that occur in the typical sequencing of a Lyme disease infection. The prevalence of arrows in this influence diagram demonstrates the addition of multiple paths to infection. The rectangles add a significant layer of risk messaging and show key points where communications can be crafted to address audience knowledge gaps developed from the interviews process.

As noted above, Morgan and colleagues have successfully used this approach in the development of multiple public health documents that have received widespread distribution. This process and the development of the materials as described in his book on the subject are costly and time consuming. Morgan estimated that a "crude" version of the process would take at least one calendar month with the full-time efforts of several individuals. Full efforts, completed to scientific standards of quality, take three months to a year and costs upwards of $100,000 (in 2002 dollars). While daunting, Morgan argues that "the attractiveness of this investment depends on the consequences of misunderstanding" (Morgan *et al.* 2002, p. 31).

However daunting, it can certainly be argued that the importance of determining in advance where the gaps in knowledge lie within the targeted audience serves a significantly useful purpose, and efforts aimed at discovering this are worth at least some effort of the part of communicators. Constructing a simplified version of an influence diagram and performing a shortened version of the interview process can help achieve

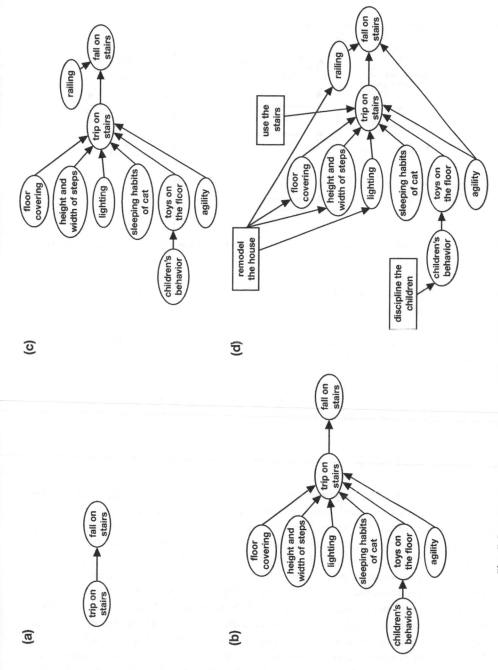

Fig. 5.2. Influence Diagram Showing Risk of Falling Downs Stairs at Home (Morgan et al. 2002)

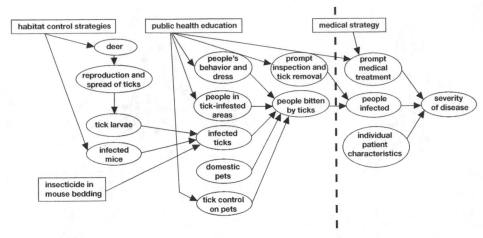

Fig. 5.3. Influence Diagram Showing Risk of Contracting Lyme Disease (Morgan *et al.* 2002)

this goal and provide a wealth of information helpful when crafting messages to be delivered verbally or shorter written messages. Morgan agrees by saying:

> A full influence diagram, developed through repeated iterations with multiple experts, can be a daunting place to begin the study of a problem. Moreover, even a rough approximation will provide much of the guidance needed for creating effective risk communication. These diagrams do not require the detail and precision required for performing quantitative analyses. Nonetheless, pushing the analyses as far as possible helps to refine thinking about a risk ((Morgan *et al.* 2002, p. 43).

MESSAGE MAPPING

As valuable as influence diagrams and the process used to create them can be, the technique requires more skill, time, and resources than is likely to be available to the typical risk communication team. The influence diagram process is also more effective when used for crafting lengthy messages in written form such as public health brochures and educational materials intended for mass distribution. Message mapping presents a more pragmatic alternative and is also a more effective technique for crafting verbal, and sometimes written, messages. It was created by Vincent Covello in the early 1990s and became a vital public health tool of the U.S. Centers for Disease Control (CDC) following the anthrax attacks of 2001. Since then, the CDC and other state and local agencies have conducted message mapping workshops for a variety of public health crises including smallpox and various water security emergencies. The technique, its benefits, and process have been well described in literature by Covello (U.S. EPA 2007; Covello 2002).

The foundations of the process address several key theories of risk communication including the risk perception model and the mental noise model (both discussed in greater detail in Chapter 3). Risk perception factors that create significant amount of mental noise include those out of the control of the audience, are involuntary, are inescapable, and are exotic. Injuries that are dreaded and significant amounts of uncertainty are also prone to creating high levels of worry and anxiety. The message mapping process provides solutions to overcoming these barriers by developing a "limited number of key messages that are brief, credible and clearly understandable" (U.S. EPA 2007, p. 2-1).

The effective use of message mapping allows organizations to identify key audiences in the early stages of message development, determine their key questions and concerns, and fill the gap between what an audience knows and what the organization wants the audience to know in order to increase the audience's ability to make informed decisions.

The expected outcomes experienced by organizations that use this process are numerous, but most important is the ability of the organization to develop a vetted repository of messages in advance of a crisis, thus allowing them to be able to quickly get information out that is accurate, requiring only a limited amount of time to refresh content or adjust to the unique circumstances of the particular crisis. In addition, in situations when there are multiple communicators from the same or related organizations, the ability to speak with one voice is increased when the message content is pre-developed and agreed upon. For communicators, message maps provide a simple visual aid to help guide them in communication events, thus reducing the chances of making an inappropriate or inaccurate statement or forgetting a key point. Finally, the messages developed are able to condense key information into brief messages of limited words, providing excellent sound bites and quotes for the media.

A previously completed message map that was used to develop messages to help the public and media differentiate between seasonal flu and pandemic flu appears below.

The process of creating message maps involves seven steps, summarized below (U.S. EPA 2007; Covello 2002):

1. **Identifying potential stakeholders.** Each crisis will have a unique set of stakeholders and include external publics and well as internal ones. A comprehensive list can number well over 70 identified stakeholders; however, most crises have a smaller list, and if too large, the stakeholders can be subdivided according to the types of concerns and questions they may have. It is also key to note that not every stakeholder needs its own set of maps; many overlaps occur. Some examples of typical stakeholders include emergency response personnel, internal workforces, law enforcement personnel, medical practitioners, public health officials, publics at risk, scientific leaders, suppliers/vendors, and victim's families.

2. **Identify potential stakeholder questions and concerns.** This list will also be different for each stakeholder group that is identified, although some common groups can be put together. It is critical to the process to brainstorm in this step in order to develop the most comprehensive list possible; future steps will allow

for a winnowing down to the most critical questions. Covello suggests a framework for listing the questions/concerns be used that includes:

- Overarching questions—What essential information do people really need to know?
- Informational questions—What details about the crisis do people need to know?
- Challenging questions—Why should the public trust what you are saying?

Developing questions can also be a time-consuming step and if rushed can distort the actual message development. Covello recommends using research to explore and confirm the list. Some possible sources of information include media and website content analysis; review of public documents such as hearing records and legislative transcripts; reviews of organizational complaints, logs, and hotline calls; and focus groups or surveys. [Covello also provides a list of the 77 most frequently asked questions by the media (U.S. EPA 2007, p. 2-5).]

TABLE 5.3. Message Map for Communicating Differences between Pandemic Influenza and Seasonal Influenza[a]

Pre-Event Risk Communication Message Map for Pandemic Influenza

Stakeholder: Public and Media
Question or Concern: How is pandemic influenza different from seasonal flu?

Key Message 1:	**Key Message 2:**	**Key Message 3:**
Pandemic influenza is caused by an influenza virus that is new to people.	The timing of an influenza pandemic is difficult to predict.	An influenza pandemic is likely to be more severe than seasonal flu.
Supporting Fact 1-1: Seasonal flu is caused by viruses that are already among people.	**Supporting Fact 2-1:** Seasonal flu occurs every year, usually during winter. Seasonal flu occurs every year, usually during winter.	**Supporting Fact 3-1:** Pandemic influenza is likely to affect more people than seasonal flu.
Supporting Fact 1-2: Pandemic influenza may begin with an existing influenza virus that has changed.	**Supporting Fact 2-2:** Pandemic influenza has happened about 30 times in recorded history.	**Supporting Fact 3-2:** Pandemic influenza could severely affect a broader set of the population, including young adults.
Supporting Fact 1-3: Fewer people would be immune to a new influenza virus.	**Supporting Fact 2-3:** An influenza pandemic could last longer than the typical flu season.	**Supporting Fact 3-3:** A severe pandemic could change daily life for a time, including limitations on travel and public gatherings.

[a]Covello 2008

3. **Analyze questions to identify common sets of concerns.** If the brainstorming process of the previous step was effective, a lengthy list of questions and concerns emerges with limited obvious patterns. However, upon further study, most lists can be further categorized into 15 to 25 overarching areas of concern. Some of the most common to look for include accountability, basic information and details, duration, effects on health, legal/regulatory, safety, trust/credibility, and quality of life. [Covello also provides a list of some of the most likely categories (U.S. EPA 2007, p. 2-6).]

A fairly simple approach to categorizing the list of concerns is to simply construct a matrix that lists the major categories and the major stakeholders as viable and using simple hash marks to identify the combination of categories and stakeholders that appear most often. This process will be very useful in subsequent steps when the actual messages need to be constructed with a very limited number of words.

4. **Develop key messages.** Because each concern or question grouping will have its own map of three key messages, this step can be very time consuming. Recall, however, that the maps developed become a permanent part of an organization's repository and once completed are done except for occasional review and revision. The use of a team is necessary for this step as well and should include representatives from the organization including subject matter experts, communication specialists, policy and legal experts, and management, all led by a skilled facilitator.

Covello recommends that the messages have a very tight structure and limited content, matching typical media sound bites. Therefore, the message length should be no longer than 27 words and/or be able to be read in nine seconds. He also suggests that those messages containing "compassion, conviction, and optimism" are the ones most likely to be used by media and repeated by the stakeholders (U.S. EPA 2007, p. 2-7). In addition, in order to develop a number of messages that can be effectively delivered by communicators, Covello suggest three key messages to address each concern/question. This leads to the 29/9/3 template he advocates.

5. **Develop supporting facts.** For each of the three key messages developed in the map, three separate supporting facts should be provided by the available research and literature. These supporting factors provide additional information for the communicator and lend credibility to the message. Like key messages, supporting facts should be research based, using the same sources noted in Step. 2.

6. **Test and practice messages.** Standardized procedures for message testing include asking subject matter experts not directly involved in the communication event to validate the message content and then participating in a practice session by delivering specific messages with groups that are representative of the key characteristics of the eventual intended audiences. Feedback should be sought from the audiences in order to opportunities to revise as necessary.

7. **Delivery of maps through appropriate channels.** The use of trained communicators is essential in this step and is addressed in greater detail in Chapter 7. Typical channels include various governmental agencies and the media through press conferences and releases, informational forums, internal staff meetings, community meetings, and written content that might appear on a website, brochure, or FAQ (frequently asked questions) sheet.

In addition to the information presented above, a number of guiding principles are helpful to keep in mind when developing message maps. These principles help understand how to craft the message content so that the audience is more likely to understand and respond appropriately to the message. They also focus on the audience's ability to understand the message in terms of relational content and the potential problems produced by high stress levels (U.S. EPA 2007).

- **The rule of 3.** Research has shown that when mental noise is high, the ability of an audience to process large quantities of information and/or specific messages is limited. Keeping the number of messages to three in high-stress situations appears to be the optimum number. Under less stressful circumstances, audiences typically can process seven separate messages effectively.
- **Primacy/recency.** In high-stress situations, audiences typically process what they hear first and last more effectively than information presented in between. Therefore, risk communicators should strive to assure that the most important information is presented in those positions.
- **AGL-4.** A typical audience in an industrialized country not under stress can process messages at the 10th- to 12th-grade level. The effects of stress are believed to reduce the comprehension level by four grade levels. Therefore, crisis messages for industrialized audiences should be at the 6th- through 8th-grade level. (Obviously, messages for nonindustrialized populations would be based upon their expected comprehension level.)
- **Triple T model.** An axiom of nearly every communication situation from classroom lectures to safety training follows the formula of: (1) tell the audience members what you are going to tell them; (2) tell them; (3) tell them what you just told them. Crisis communication events are no different. This formula also addresses and helps reduce the effect of mental noise by increasing the chances that repetition of the content will overcome the barriers produced.
- **1N = 3P.** As was discussed in Chapter 3, the negative dominance model suggests that the ability to hear positive messages in a crisis situation is generally undercut by negative messages. In order to increase the chances that positive messages will be heard, they need to be delivered in a 3:1 ratio against negative messages. This principle also cautions the communicator to strongly refrain from using highly negative words in any message such as "no," "never," "not," "nothing," and "none."
- **Anticipate, prepare, practice.** As was already discussed in Chapter 3 and will be further elaborated upon in Chapter 7, choosing a spokesperson requires an

understanding of the difficulties posed by the communication event. Even more so, crisis communicators need to spend time in preparation for the event, even going so far as to rehearse with a mock audience. At the very least, anticipating possible questions that will follow the delivery of a message and knowing how to answer them is essential. As noted above, the prepared message map is an effective visual tool for communicators and aids in helping them to "stay on message," especially when the audience is angry or highly emotional.

- **Cite third parties.** As has been reinforced numerous times throughout this text, trust and credibility of the communicator are keys to any successful message delivery event. Knowing the audience in advance of the event will help identify the trust and credibility level and determine whether or not the need to cite credible third parties in the message will help. It also goes without saying that ensuring the validity of third parties and the audience's acceptance of them as credible sources needs to occur before the message event.

- **Use graphics and other visual aids.** Chapter 7 will discuss the importance of and the use of graphics and other visual aids during the message delivery; suffice it to say at this point that visuals provide value by increasing the chance that the message will be understood by both reinforcing the content and also by sending the information content to a different part of the audience members' brains, which aids in comprehension and retention. In addition, visuals can help overcome mental noise because they are often simpler to process.

REFERENCES

Coombs, W.T. 1999. *Ongoing Crisis Communications: Planning, Managing, and Responding.* Thousand Oaks, CA: Sage Publications, Inc.

Covello, V. 2002. "Message Mapping, Risk and Crisis Communication." Invited paper presented at World Health Conference on Bio-Terrorism and Risk Communication, Geneva, Switzerland, October 1.

Covello, V. 2008. "Risk Communication: Principles, Tools and Techniques." Posted online at http://www.maqweb.org/techbriefs/tb49riskcomm.shtml on February 25, 2008. Accessed on October 29, 2009.

Covello, V., R. Peters, J. Wojtecki, and R. Hyde. 2001. "Risk Communication, the West Nile Virus Epidemic, and Bioterrorism: Responding to the Communication Challenges Posed by the Intentional or Unintentional Release of a Pathogen in an Urban Setting." *Journal of Urban Health* 78(2):382–391.

Covello, V.T. *et al.* 1988. "Seven Cardinal Rules of Risk Communication." U.S. Environmental Protection Agency, Washington, D.C. Distributed by Office of Public Liaison [Pennsylvania] Dept. of Environmental Resources.

Lundgren, R.E. and A.H. McMakin, A.H. 2004. *Risk Communication: A Handbook for Communicating Environmental, Safety, and Health Risks*, 3rd ed. Columbus, OH: Battelle Press.

Morgan, M.G., B. Fishhoff, A. Bostrom, and C.J. Atman. 2002. *Risk Communication: A Mental Models Approach.* New York: Cambridge University Press.

Peters, R.G., V.T. Covello, and D.B. McCallum. 1997. "The Determinants of Trust and Credibility in Environmental Risk Communication: An Empirical Study." *Risk Analysis* 17(1):43–54.

Sandman, P. 1991. "Twelve Principle Outrage Components." Posted online at http://www.psandman.com/handouts/sand58.pdf. Accessed on January 11, 2010.

Sandman, P. 1995a. "Assessing Stakeholder Motives: The Three Main Reasons for Making Demands." Posted online at http://www.petersandman.com/handouts/sand3.pdf. Accessed on March 19, 2010.

Sandman, P. 1995b. "The Nature of Outrage." Posted online at http://www.petersandman.com/handouts/sand31.pdf. Accessed on March 19, 2010.

Sandman, P. 2003. "Four Kinds of Risk Communication." Posted online at http://www.petersandman.com/col/4kind-1.htm on April 11, 2003. Accessed on January 29, 2010.

Sandman, P. 2004a. "Crisis Communication: A Very Quick Introduction." Posted online at http://www.petersandman.com/col/crisis.html on April 15, 2004. Accessed on April 15, 2008.

Sandman, P. 2004b. "Worst Case Scenarios." Posted online at http://www.petersandman.com/col/birdflu.htm on August 28, 2004. Accessed on April 15, 2008.

Sandman, P. 2007. "Empathy in Risk Communication." Posted online at http://www.petersandman.com/col/empathy.htm on July 29, 2007. Accessed on January 11, 2010.

U.S. Department of Health and Human Services. 2006. "Communicating in a Crisis: Risk Communication Guidelines for Public Officials." Washington, D.C.

U.S. Environmental Protection Agency. 2007. "Effective Risk and Crisis Communication During Water Security Emergencies." EPA 600-R07-027.

U.S. Federal Emergency Management Agency. 2004. "Are You Ready?" Washington, D.C.: Department of Homeland Security.

6

DELIVERING THE MESSAGE WHILE AVOIDING COMMON MISTAKES

This chapter begins with the assumption that the message has been crafted and is ready for delivery, either verbally or in a written form. While much has been said in previous chapters regarding the importance of content, the significance of appropriate delivery cannot be understated. From the setting to the tone of the words to the communicator's nonverbal gestures, understanding and applying delivery concepts is vital to a successful communication event. The first part of this chapter will be devoted to an elucidation of those concepts while the latter portions will identify typical errors made by communicators and organizations while delivering the messages. The chapter will then conclude with a discussion of evaluating the communication event.

MESSAGE DELIVERY TEMPLATES

Chapter 5 previously noted several key message crafting templates and principles, but there are also several key message delivery templates and principles generally shared by most experts in the field, all of which enhance the success of the message in achieving its intended goals and objectives while either building or rebuilding the trust and credibility of the communicator. These are also techniques that can be used in varying

Risk and Crisis Communications: Methods and Messages, First Edition. Pamela (Ferrante) Walaski.
© 2011 John Wiley & Sons, Inc. Published 2011 by John Wiley & Sons, Inc.

situations. The following list comes from a U.S. Department of Environmental Protect Agency document created to assist spokespersons during water emergencies, but the applicability of its points is universal, regardless of the type of message or event it is created to address (U.S. EPA 2007).

1. **Bridging templates.** These are effective statements that can be made by a spokesperson that assist in returning to key message points in order to emphasize them or to redirect the communication event when it has strayed off of its original objectives. Some examples include:

   • "However, the real issue here is. . . ."

   • "Let me put this all in perspective by saying that . . ."

   • "What matters most in this situations is . . ."

   • "Before we leave the subject, let me add that . . ."

2. **IDK templates (I don't know).** These templates are useful when the communicator does not know the answer, cannot answer, or is not a good source for the answer, even if they believe they know it. Use of this template involves several steps that include:

   • Repeating the question and deleting any negatives in it.

   • Acknowledging the inability to answer the question. Say something to the effect of, "I wish I knew the answer to that," or simply, "I don't know the answer to that."

   • Explain why the communicator cannot answer.

   • Provide a method for getting the answer and a timeline.

   • Bridge to what the communicator can say. Sentences such as "What I can say is . . ." work well.

3. **Guarantee templates.** It is almost never a good idea to guarantee anything as a spokesperson, but often an audience or individual audience member asks for just that. This template provides a way to answer the request without guaranteeing. Again, there are recommended steps:

   • Indicate that the question is about the future. ("You've asked me to guarantee something in the future.")

   • Indicate that the past or present can help predict the future. ("The best way I know to discuss the future is to look at the current and previous situations or events.")

   • Bridge to known facts or actions. ("And what we know from yesterday is . . .")

4. **"What-if" template.** Questions from audience members often begin with "what if" or some variation of that framework. The audience appears to looking for guarantees or assurances that may not be able to be given for a variety of reasons; therefore, the steps for this template are similar to questions asking for guarantees and include:

- Repeat the question without the negatives. ("You've asked me what might happen if. . . .")
- Bridge to what is. ("I believe it is valuable to talk about what facts we know now . . .")
- State what is known factually. ("And what we know is . . .")

5. **False allegation template.** In many situations audience members will make deliberately false allegations or allegations that are deliberately embellished to prove a point. They may also ask hostile or critical questions. The steps to address these highly charged emotional situations are as follows:

- Repeat the question without the negative, or use more neutral language. ("You've raised a serious question about X . . .")
- Indicate that the issue is important. ("X is an important issue to discuss . . .")
- Indicate what has been done/will be done to address the issue. ("We have done the following to address X . . .")

Table 6.1 below provides some quick and general usage bridging statements.

Additional "pitfalls" in message delivery are offered by the EPA, this time in a document that is intended to provide a general framework for all types of crisis communications but also provides more specific ideas for communicator during a verbal presentation (U.S. EPA 2005).

1. **Avoid abstractions and jargon.** Using abstractions or jargon assumes a common level of knowledge and understanding about technical information between the communicator and the audience. If audience members are confused about what the communicator is saying, they are likely to become frustrated or angry during the communication event, particularly in high-stress situations.

TABLE 6.1. Effective Bridging Statements[a]

Ten More Effective Bridging Statements	
1	And what's most important to know/remember is . . .
2	With this in mind if we take a look back . . .
3	What this all boils down to is . . .
4	And as I said before . . .
5	Let me point out again that . . .
6	Another thing to remember is . . .
7	Let me emphasize again . . .
8	And that reminds me . . .
9	Here's the real issue . . .
10	It is true that . . . but it is also true that . . .

[a]Hyer and Covello 2005

2. **Don't attack or blame.** Responding defensively to an attack by the audience only fuels a confrontation. Delivering messages that demonstrate that the organization is willing to accept some of the blame for the situation will help the audience see that the organization is being fair and honest with the audience.

3. **Remember to convey positive nonverbal messages.** An open stance while communicating and/or listening is critical. Open stances include arms at the side or forearms resting on a podium, while closed stances would include crossed arms. Body language that conveys anger, frustration, or smugness will overwhelm even the most carefully crafted message. Examples of these include scowling, eye rolling, sighing, or lack of eye contact.

4. **Avoid guarantees, promises, and speculation.** Emphasize what is likely to happen and what progress has been made in the communication event, not on what is not yet known or accomplished. Focus on what the organization is willing to try and deliver, without promising. Don't try to predict the future.

5. **Don't go for the joke.** In times of stress and crisis, humor is most likely going to fail, or worse, backfire. It occasionally can be directed at the communicator, but even those attempts are fraught with difficulties. Humor is best left to those who have significant experience in communication events and only in some unique circumstances.

6. **Don't go on and on.** Aim for no more than 15 minutes of presentation in a communication event without stopping for questions, dialogue, or a change in speaker or focus. Presentations longer than 15 minutes increase the chance that the audience will tune out. Make sure to leave plenty of time for questions if the event is designed for that.

7. **Don't repeat negative allegations.** It is okay to refute them—succinctly and with limited detail—but be sure it is done only in ways that don't give the allegations credibility.

8. **Avoid negative words and phrases.** As noted in Chapter 5, the use of words like "no" and "never" should be avoided. Remember that every negative message should be countered with at least three positive messages to overcome the tendency to place a greater weight and credibility on negative messages.

9. **Communicators are representatives of their organization.** Always avoid the use of the word "I" in messages; use "we," "our company," or use the specific name of the organization. Using "I" conveys an authority on the messages that does not exist and does not portray the organization as being unified on the message content or position.

10. **Communicators are never off the record.** Everything that is said in a communication event is a part of the public record, regardless of whether or not the words are being recorded.

11. **Don't rely on words alone.** Visuals are important and help explain key points. They also help audience members retain information because it is conveyed to them in two different formats. Finally, it helps take away from the technical nature of the presentation and helps the audience listen and absorb the content. (More on this topic later in this chapter.)

12. **Limit the use of statistics.** Statistics are helpful to emphasize or clarify a point. If they are used as the bulk of the presentation, most audiences will tune out to the more important points that need to be made and heard. If statistics are presented, complement the words with graphics that help illustrate the concepts.

In addition to the above guidelines, Lundgren and McMakin (2004) offer a few more:

- If the best spokesperson for the organization has little or no public speaking experience, it is best to either attempt to identify someone who does, or provide the training to the chosen communicator well in advance of the communication event. There are numerous public relations courses, seminars, and individualized training available that specialize in risk and crisis communications. Also helpful are the more traditional routes used by those who are learning to speak in generalized public situations like Toastmasters or the Dale Carnegie Foundation. Training should emphasize basic public speaking skills as well as learning how to deal with an emotional audience (angry, mistrustful, or fearful). It should also provide opportunities to understand how to deal with the media.
- It is essential that the communicator spends some advance time practicing the presentation. Sometimes it is helpful for the communicator to deliver messages to a receptive audience at first before progressing to a more difficult one. In addition, the more emotional the audience is expected to be, the more practice should occur. If at all possible, the practice session should mimic the expected room setting. Consider having coworkers act as audience members to role play reactions and ask questions. A thoughtful audience profile as was discussed in Chapter 4 will aid in these types of preparations.
- Choose settings wisely. Look for communication locations that are comfortable, accessible, and in a neutral location. And consider the staging of setting. A raised platform conveys authority and importance, which creates the opposite effect in a communication event where feedback and dialogue is sought. In addition, if an organization is already mistrusted by the audience, elevating the communicator may further engender mistrust by the audience.
- Don't forget to consider the setting's acoustics and lighting. Make sure audience members in the back of the room can hear the communicator and see any visuals. When using presentations that are projected on a screen, dim the lights somewhat, but be careful not to make the room too dark; audience members won't be able to see the speaker or each other and may even fall asleep or begin side conversations. Don't forget to consider the temperature of the room. Being able to adjust it easily or have someone nearby who can do so is important.
- Communicators should think about their clothing choices. Formal presentations call for formal outfits such as suits or dress shirts and pants. Communicators in informal presentations can safely wear jeans and casual shirts. Outdoor communication events in the middle of a natural disaster crisis such as a hurricane

or tornado would be an acceptable place for rolled up sleeves or even t-shirts. Some organizations require the donning of shirts with logos.

- Make sure the communicator knows how to use the audiovisual equipment or has assistance nearby from someone who does. Presentations that have to be stopped to load slides or videos or adjust settings are distracting and frustrating to the audience. The momentum is lost and may be hard to regain. It also makes the organization appear incompetent.

In Chapter 3, significant discussion outlined Covello's risk perception model, which identifies 15 factors that influence the public's perception of risk and therefore its understanding and response to risk and crisis messages. Some of the key factors discussed in that chapter included voluntariness, controllability, familiarity, fairness, and trust. These concepts also provide a sound understanding of delivery strategies as evidence in Table 6.2 presented here.

THE USE OF VISUALS IN A COMMUNICATION EVENT

Adults, who make up the majority of communication event audience members, learn and retain information in a variety of ways, although most have a preferred style from among the options of hearing, seeing, and doing. As communication events require the discussion of more technical information; involve information that an audience is unfamiliar with; or involve situations where emotions such as anger, fear, or mistrust are high, the ability to comprehend solely with one style of learning decreases. In typical risk and crisis communication situations, hearing is the preferred method of learning. However, the ability to comprehend and retain information decreases as the level of technicality increases.

For these reasons, it is generally a good idea to ensure that verbal presentations are accompanied with some sort of a visual. Table 6.3 provides some illustrations on how to match the type of risk information being presented with the visual options that work best. For example, if the information about the risk being conveyed involves data and numerical ratings of risk, numerical visuals or charts are the best choice. The effects of the risk that are visual and can be seen are best presented through the use of photographs or drawn illustrations.

The U.S. Department of Health and Human Services (2006) has developed guidelines and concepts that are important in understanding how to effectively use visuals during communication events. These include:

- Ensuring that the visual can be seen by all members of the audience. This often eliminates posters and other smaller visuals in larger groups. If the visual can be recreated on a one-page handout (front and back), these smaller front-of-the-room visuals might still work. Illustrations that can be projected on large elevated screens will work best in most situations.
- Using color to enhance visual presentations. Placing a visual on a colored background if the information being presented is textual in nature will help

TABLE 6.2. Strategies for Delivering Targeted Messages[a]

To communicate voluntariness, deliver messages that:

- Make the risk more voluntary
- Encourage public dialogue by using two-way communication channels
- Ask permission
- Ask for informed consent

To communicate controllability, deliver messages that:

- Identify things for people to do (for example, precautions, preventative actions, and treatments)
- Indicate your willingness to cooperate and share authority and responsibility with others
- Give important roles and responsibilities to others
- Tell people how to recognize problems or symptoms
- Tell people how and where to get further information

To communicate familiarity, deliver messages that:

- Use analogies to make the unfamiliar familiar
- Have a strong visual content
- Describe means for exploring issues in greater depth

To communicate fairness, deliver messages that:

- Cite credible third parties
- Cite credible sources for further information
- Acknowledge that there are other points of view
- Indicate a willingness to be held accountable
- Describe achievements
- Indicate conformance with the highest professional, scientific, and ethical standards
- Cite scientific research (be prepared to point to specific published studies)
- Describe the review, approval, and advisory processes
- Identify the partners working with you
- Indicate your willingness you share the risk ("do unto others only that which you would be willing to do unto yourself or your loved ones")

[a]Hyer and Covello 2005, p. 39

as long as the background color is pastel or a light shade and the text is bold enough to be read. Use different colors to enhance important points or to contrast more than one point on an individual visual, but avoid red as a font color for text.

- Creating a visual that, while not intended to, can stand alone if it is a depiction of the information being presented. Once the visual is presented to the audience, its attention to spoken words will decrease so the visual needs to be able to stand

TABLE 6.3. Options for Portraying Various Aspects of Risk Visually[a]

Risk Information	Options for Visual Format
The risk and its effects	If the effects of the risk can be seen (such as visible health effects, effect on pants and foods, etc.), depict them in a photo or illustration to help people identify the risk. Also consider showing conditions leading to or indicating a risk, such as blocked fire doors in an industrial plant, high-power lines for electromagnetic fields, or people demonstrating unhealthy or unsafe behaviors and their consequences.
Size and significance of the risk	Show the risk in the affected population, using numbers or charts (such as a line graph or bar chart). Show the risk over time as it increases or decreases. Compare judiciously with other similar risks or show relative magnitudes. Consider including a recommended "action" level—a point at which people may want to take action to mitigate the risk.
Likelihood of risk for specific people	Show probabilities and uncertainties for various conditions. Consider phrasing as "X in Y chances of occurring" under certain conditions. Tables, charts, and graphs can show various risk levels for various situations. "If/then" flow charts can help people walk themselves through the risk probabilities.
Change over time	Use graphs, charts, or pictograms (small pictures representing the risk) to indicate trends over time. Consider several different representations if many variables are involved, such as conditions that change the risk over time.
Alternatives to the risk, with corresponding benefits and dangers	Compare alternatives with pros and cons of each. Consider using tables if there are shared variables among the alternative and the alternatives are being compared in some way (costs, environmental effects, health effects, etc.). If the alternatives are not easily comparable, use formats that do not invite comparison on the same scales.

[a]Lundgren and McMakin 2004

on its own for a period of time until the audience's attention is brought back to the speaker.

- Ensuring that each visual contains only one major idea or key concept. If more than one concept is required to be communicated, several separate visuals that are connected by consistent colors and presentation styles may be necessary. Remember to eliminate unnecessary clutter from the visual—stick to the key points.

- Limiting the number of words on any one visual. A good rule of thumb is no more than six words to a line and no more than ten lines per visual. More words cause the audience to stop listening to the communicator and read the visual for long periods of time. Visuals are intended to enhance and illustrate a point, not make it.

- Using professionally created visuals only, not hand drawn ones.

DELIVERING THE MESSAGE IN THE AGE OF THE INTERNET

While much of the discussion regarding communication events thus far has focused on the traditional venues such as public meetings, news reports, press conferences, internal workforce trainings, and similar face-to-face contacts, the past 10 to 15 years has seen the development of new and previously unavailable message delivery options, most often utilizing the Internet as their platform. They include the websites of various organizations such as public health departments and news providers, as well as social networking pages like Facebook, Myspace, and Twitter, and individual blogs. In fact, the H1N1 pandemic was reported to be "the first pandemic with a 'blogosphere' and other rapid social messaging tools that challenged conventional public health communication" (Leung and Nicoll 2010). This significant change in the ways in which audiences communicate with each other will require organizations to adapt their message methods as well if they wish to reach the audience successfully.

In evaluating the various types of message delivery mechanisms, organizations would be well served to begin to evaluate and utilize the newer technologies to reach targeted audience segments in addition to the more traditional methods. Chapter 4 discussed the process of audience profiling as a means of developing targeted audience segments based on demographic data. This data can then be used to determine which message mechanism is most likely to be successful.

It is not surprising that younger audience segments are the ones more likely to receive messages from newer technologies from simple text messages to more complicated e-mails. And as the younger audiences age the mechanisms that appear today to be "newer" will become more commonplace. Table 6.4 lists some of the various mechanisms and technologies, identifying some of the pros and cons of each one. While it focuses mostly on newer message delivery systems, it does also include traditional methods. The critical aspect for an organization is to understand its audience and the circumstances under which it will need to deliver messages in order to select the best-suited methods.

A study undertaken regarding the success of the various Internet message attempts during the H1N1 pandemic indicated the traditional risk communication community has much to learn to take advantage of this delivery method (Gesualdo et al. 2010). The study authors conducted a search in August 2009 for "swine flu" (the name by which H1N1 was originally referred), accessing three of the most popular search engines in Australia, Canada, the United Kingdom, and the United States. They were looking for sites that provided information regarding the World Health Organization's (WHO) recommendations for flu prevention. (The flu was originally identified in Mexico City in April 2009, declared a pandemic by WHO in June and was nearing the height of its confirmed cases worldwide by August. See Chapter 10 for an in-depth case study on messaging during this event.) Using three popular search engines (Google, Yahoo, and MSN-Bing), 147 websites were located. The study authors explored each of the sites using all of the links, using no more than four clicks from the search engine Web page and looked for specific information on how to prevent spread of H1N1, specifically information on WHO-recommended actions. (See Table 6.5 for a list of the specific recommendations searched.) The authors

TABLE 6.4. Comparison of Message Delivery Mechanisms

Mechanism	Pros	Cons
Telephone text messages	• Percentage of cell phone use nears 85 percent • Sender has control of initial message • Majority are delivered quickly to user • Is an active communication method	• Text message capacity is not a standard part of calling plans • Costs to send are high • Some delivery failures and delays beyond control of sender • Receiver has to sign up for service and provide updated phone numbers if they change
Blast e-mails	• Sender has control of initial message • Messages can be sent to specific target groups • Messages can be sent quickly • Easy to update original messages	• Receiver must be online or have phone with e-mail capacity • Runs on electrical infrastructure • Receiver has to sign up for service and provide updated e-mail addresses if they change
Reverse 911: blast calls	• Receivers can access remotely • Interface for message delivery is secure • Inbound phone lines can be seized in order to provide additional capacity • Phone numbers are automatically updated when receiver changes	• An expensive method to utilize • Receiver has to sign up for service and provide updated phone numbers if they change
Voiceover IP (Internet protocol)	• One message can be transmitted over multiple outlets • Cost is low compared to usage • Allow sender to prioritize phone traffic • Works if land lines are not available • Allows an interactive survey tool	• Upfront cost is high • Can be challenging to implement without internal technical capability • Must have a stable IT network • Receivers must access their e-mail to obtain messages • Runs on electrical infrastructure
Hotlines	• Provides a consistent controlled message • Gives sender feedback from receivers • Keeps the use of main phone line manageable • Some receivers prefer to hear a human voice so use of interactive operators is appealing	• Calls in can get lengthy • Requires intensive training of operators • When messages change, operators need to be retrained • If message-only capacity, no human connection to receiver
Crisis website	• Sender controls information posted • Can post a substantial amount of information • Information can be quickly updated • Information can be retained and archived	• Passive communication delivery • Traffic to site can overload system • Runs on electrical infrastructure • Receivers must have internet access

TABLE 6.4. (*Continued*)

Mechanism	Pros	Cons
Public address system (internal)	• Rapid message delivery • Consistent message delivery • Targeted message delivery • Good for simple messages that provide instruction on action to be taken	• No immediate feedback • Some receivers feel it invades privacy • Can only hear if near speaker (better systems are being developed)
Outdoor messages	• Receiver controls the message • Covers a broad area • Inexpensive to set up	• Receivers may see as impersonal • Receiver must be nearby to get message • Runs on electrical infrastructure
Computer pop-ups	• Receiver controls the message • Limited cost to implement	• Receivers can easily ignore request to access message • Receivers must be at computer to obtain message • Runs on electrical infrastructure
RSS feeds	• Sender controls the message • Messages are desired by receiver via sign-up • No cost to sender	• Passive communication • Receivers may not access in a timely manner
Social media	• Nearly two-thirds of Americans visit social media sites regularly. • Quick to set up • Easy to access and send messages from mobile phone • Receiver controls initial message	• Receiver has to build a group to follow • Once message is delivered, additional messages are not controlled by sender • Sites frequently change processes, set-up, and rules for posting • Runs on electrical infrastructure
Blogs	• Read regularly by 30 million people in U.S. • Sender controls the message • Allows receiver to keep a record of messages sent • Inexpensive to set up and maintain • Receiver can elicit feedback • Media use them to obtain information for news stories	• Must take the time to develop a following • Must respond to feedback • Sender does not control feedback • Can be easily overrun by those who oppose the sender • Receiver may not get message in a timely manner

TABLE 6.5. Internet Availability of Risk Communication Messages During H1N1 Pandemic[a]

Recommendation	Percentage of Sites Reporting
Hand cleaning	78.91%
If you are sick, stay home	77.55%
Dispose of tissue after using it	75.51%
Cover mouth and nose with tissue when coughing or sneezing	74.83%
Avoid touching mouth and nose	66.67%
Wash your hands after sneezing or coughing	65.31%
If you are sick, avoid contact with other people	62.59%
Avoid close contact with sick people	61.90%
Reduce time spent in crowds	59.86%
Cover mouth and nose with elbow if no tissue available	34.01%
Contact a health professional before traveling to a facility*	31.97%
Practice good health habits	29.25%
Inform family and friends if you are sick	17.69%

*If able to be treated at home, avoid overcrowding waiting rooms with sick people and further spreading the flu.

[a]Gesualdo et al. 2010

credited the site only when the information accessed was consistent with the WHO recommendation.

The majority of the sites reviewed were from public health agencies (37 percent or 54 sites) and news providers (40 percent or 59 sites). Other sites included those from universities, hospitals, research institutes, drug companies, and independent blogs. Table 6.5 lists some of the findings of the study.

While some of the WHO recommendations were located in fairly high percentages, many were not, and several key messages were well under 50 percent. The study results are further restricted because the study authors reviewed search engines from only four countries, although they represent the countries where the Internet is most widely used. The study authors expressed concern that an Internet-based risk communication message delivery system was missing many key messages and that those messages were not easily located once a user entered the site.

The study authors point out that providing risk communication messages on the Internet is a valuable method because it achieves two key goals: It reaches a particular segment of the audience that may have been unreached in the past because that segment was not as likely to participate in the more common face-to-face message events (e.g., young people). And its ability to reach large portions of the audience quickly makes it an ideal place for them to retrieve the information they need. The authors conclude by saying:

> Thus, in order to achieve a change of behavior in the general public and favor the implementation of such recommendations, health professionals, using communication

strategies, have informed the population through different channels, including mass-reach broadcast media advertising (television and radio advertisements or programs), print-based materials, and audiovisual materials. The role of the Internet as a platform for delivering public health interventions to specific patient groups and to the general public is constantly increasing, due in particular to its disseminating potential; the worldwide penetration of the Internet is increasing and the use of this medium for seeking health information is frequent. Moreover, the Internet potential for individual tailoring and interactivity is superior to that of other high reach delivery channels. Integration of the Internet with the more classical media, as well as with the novel, alternative formats (e-mail, interactive digital TV, SMS texts, podcasts) for social marketing can allow reaching specific groups whose characteristics are different from those of the general population.

COMMON MESSAGE DELIVERY MISTAKES AND EFFECTIVE CORRECTIONS

Even after an organization spends time outlining the purpose and objectives of the communication event, does a thorough audience profile, and crafts messages that are well thought out, the delivery of the words can obstruct the message from succeeding or even end up creating a much worse situation than existed before the communication event. No one communication event runs completely free of error. The following list provides a summary of some of the more common mistakes made by organizations and their communicators during a message delivery event, as well as those that are most likely to derail the effort.

Failing to Communicate Technical Information

In the first few communications events when an audience is not expected to have a firm grasp of technical information, it helps to lay the groundwork with a few simple concepts, phrases, and scientific terms. Those items will be the ones that are consistently used throughout all subsequent message events and can even be built upon with increasingly more technical information if warranted. The use of too many acronyms and jargon can be confusing, but a few selected ones that are well explained in the early message events can be helpful and begin to get the audience in sync with the organization. It helps the audience feel connected in some respects because of the shared bank of information. As noted above, visuals that depict complex information in simple charts or illustrations are important to aid in increasing the audience's level of understanding. Finally, using familiar frames of reference to explain technical information will help the audience members bring what they already know and connect it with what new information the communicator wants them to understand; instead of using a number to quantify the data on its own merit, place the quantity in a context that is easily understandable. For example, instead of telling the audience that an oil spill is discharging 200,000 gallons of oil per day into the water, tell it how many football fields would be filled each day by the spill (U.S. DHHS 2006).

Failing to Help the Audience Understand the Uncertainly of Most Risk Information

As has been noted in previous chapters, communication events that attempt to hide uncertainly, rather than acknowledge it, increase the audience's level of anger or mistrust or create those emotions when there may be none. Covello (2008) makes several recommendations for dealing with uncertainty:

1. Acknowledge—do not hide—uncertainty.
2. Explain that risks are often hard to assess and estimate.
3. Explain how risk estimates were obtained and by whom.
4. Announce problems and share risk information promptly, with appropriate revelations about uncertainty.
5. Tell people that what you believe either (a) is certain, (b) is nearly certain, (c) is not known, (d) may never be known, (e) is likely, (f) is unlikely, (g) is highly improbable, and (h) what can be done to reduce uncertainty.
6. Tell people that what you believe now may turn out to be wrong later.

Trying to Compare Risks

The unfortunate result of comparing risks is that different members of the audience may view the comparison differently, creating a variety of responses to the same information, since comparisons are in and of themselves subjective (Lundgren and McMakin 2004). Analogies can also come off as trivializing the risk if they are used to as a means to try and reduce anxiety in an audience, particularly with risks that may have life-threatening consequences. Ranges that show "safe" levels at one end and "dangerous" at the other can help audiences place their risk in context, but only if they personally either fall in the "safe" area or if you can give them strategies for how they (or your organization) can move out of the "dangerous" area. Standards can be used to benchmark your organization's risk against a regulatory or commonly accepted standard, but once again, placing your organization's risk below an accepted standard should be done only when the same *conditions* are in place with ranges that can be established. The use of other studies or risk assessments that have been conducted by others independent of your organization and which reinforce your conclusions or assessments are helpful only if they can be located and are close enough to your own risk assessments

Making Value Judgments about "Acceptable" Levels of Risk

While it may be helpful to place the level of risk in context with other risks or on a continuum on its own merit, the use of the term "acceptable risk" with an audience that is hostile, angry, or mistrustful can be detrimental (U.S. EPA 2005). The concept of acceptability for a non-professional audience is a value question, not a technical one. Finding ways to help audience define its acceptable level of risk can be a more effective strategy. Techniques discussed above such as using ranges that allow an audience to put its own perception of the risk in perspective not only succeed in this respect, but they also convey a sense of trust in the audience's judgment and its ability to understand technical information in relation to how it applies to its members' lives. It can also be effective to demonstrate how your organization's proposed solutions move the risk to a less dangerous level so that the audience refocuses on the improvement of the situations rather than the current risk.

Being Concerned That an Audience Will Panic

As has been discussed elsewhere (Chapter 5), it is a rare situation when an audience panics following the delivery or a risk or crisis communication message. Its reaction may not be the one that was desired, and the audience may disagree with the communicator's assessment or fail to follow recommended actions, but those responses do not constitute panic.

Using Words That Imply Negative Behaviors

Sometimes the problem relates to impromptu comments made by officials or organizational representatives; more than one politician has believed a microphone was off when it was not and paid varying prices for their indiscreet comments. Other comments are sometimes made when communicators stray from the key messages the organization wants to convey; the solution for this is to ensure that message maps are prepared in advance of any communication event and that communicators are urged to stay on point with their comments. As was also noted above, taking the time to practice the presentation or be asked potential questions by a mock audience will also help communicators stay with the intended messages.

On occasion, one word within the context of an entire well-crafted message can be detrimental. Fearn-Banks (2007) talks about the "crisis of words" that can inadvertently create hostility in the audience. During the Hurricane Katrina response in 2005, a number of media outlets and informal networks such as blogs and other public websites raised a red flag when photo captions depicting white families "finding food" or "surviving" and similar images of black families coined their activities as "looting." A comment from President George W. Bush was also seen as insensitive when he mentioned that then Louisiana Senator Trent Lott had lost "one" of his homes, but that it would be replaced with another one just as nice, while many less wealthy New Orleans residents may have lost their "only" homes forever, ones that were much smaller and less expensive than Senator Lott's. Finally, the term "refugee" was used in the early stages of the crises to depict uprooted residents in the area. Though technically correct, the term is more commonly used to describe persons fleeing their own country rather than citizens on the United States being moved to a location to live until able to return. The term "evacuee" slowly became the normative term used by many media outlets.

Responding Too Quickly or Not Quickly Enough

In a crisis, timeliness of the response by an organization is essential—not only with actions but with words. Being prepared in advance of a crisis by completing an inventory of possible crises and planning for responses for each and including message maps and other prepared documents will lessen the tendency of an organization to either say anything or say nothing, both of which are problematic responses. Coombs addresses the timeliness of a response by noting the importance of filling the need for information as the crisis begins (Coombs 1999; U.S. EPA 2007). While some organizations fear "speaking too soon," Coombs argues that the risk of saying nothing is far greater, noting that speed isn't necessarily always equated with mistakes, especially if an organization is prepared. He also notes that incorrect or unflattering information may well be what disseminated if an organization does not fill the information void. A well-cited example is that of the error made by Johnson & Johnson in the early days of the Tylenol product tampering scandal of 1982. A spokesperson stated that cyanide, which was being used to poison the capsules, was not used in any of the company's production facilities. This later turned out to be untrue, although the only source of cyanide was in a quality-control-testing laboratory that had no connection to any production facility. When the error was discovered, it was corrected by Johnson & Johnson, and the public appeared to have believed the explanation of both the error and the inability of the cyanide to come in contact with any production.

In addition to the difficulties of not providing information, silence often is equated with uncertainty and confusion within an organization, the opposite impression an organization wants to create. And unfortunately, in some cases, silence is also equated with guilt or responsibility at a level that may not be accurate, similar to the poor response of "no comment." A better alternative is to always say that "the information is not yet available," following it up with clarification on when it is to be expected and then delivering on that expectation.

As has been noted repeatedly throughout this text, the credibility of an organization (and its spokespersons) is critical to successful message delivery. Responding quickly and with authority denotes control, expertise, and credibility.

Failing to Speak with One Voice

While it may be impossible to have one spokesperson throughout the life of the incident, the "voice" of the organization must be consistent (Coombs 1999; U.S. DHHS 2006). In crises of long duration, many representatives of an organization are called upon, in part because they possess levels of knowledge and expertise that are deemed essential to the communication event, and, in part, because one spokesperson may not be able to provide all of the communication needs during a lengthy crisis or one of large-scale proportions where communication events are held frequently throughout the crisis. Regardless, the consistency of the key messages is an essential component of the success of the crisis management. As has been noted above, the use of prepared message maps or other "talking points" can be a significant factor and can provide opportunities for communicators to practice before they speak. Further, as was noted in Chapter 5 and is discussed in Chapter 7, not all people make effective communicators. Carefully evaluating and selecting spokespersons is as important as the content of the messages they deliver.

In addition, many crises involve multiple organizations, all with their own spokespeople, and each with their own message purposes and objectives as well as miscellaneous issues. Though difficult to achieve, the ideal situation is one in which multiple organizations work closely together to develop a consensus about message concepts and deliver messages that mirror each other in their overall framework. Less than ideal, but acceptable, is a situation in which multiple organizations agree at least on the major issues and deliver messages that, while not in complete harmony, at least do not contradict each other or worse yet, single out another organization for blame. Again, the lessons of Hurricane Katrina remind us that "dueling" press conferences serve only to increase public levels of hostility and damage fragile credibility and trust. More often than not, this includes the organization that believes it can deliver a message that shifts blame and responsibility onto another.

THE USE OF CONTENT ANALYSIS AND READABILITY ANALYSES

As was noted in Chapter 4, performing an audience profile prior to developing messages can help ensure that the messages that are delivered match the audience on multiple levels, including comprehension and, for written communications, grade comprehension level. It was also noted in Chapter 5, that a reducing the grade level by four in times of high stress can improve the receptiveness of the message and the ability of the audience to respond in the way envisioned by the communicator.

The use of content analysis is an effective tool for accomplishing the above objectives and is recommended for written communications with wide distribution. While typically utilized for written materials, verbal communication events to large audiences can also benefit from this technique. This is particularly true for those events when a prepared statement or speech is given, but it also can be used to check key messages for vocabulary. Finally, content analysis can be useful as a final check on the content of the words being displayed on visuals.

Content analysis involves coding and classifying information. Despite the initiation and widespread availability of software that can perform some of the analysis, the process remains labor intensive and involves training people to code the written information into well-defined categories. The strength of this process when human beings code the information lies in the definitions of the categories so that a reasonable measure of consistency is reached among multiple coders reading the same messages (Coombs 1999). While this method can be helpful for a pre-analysis of prepared messages, its usability is increased when it is applied to lengthier written communications and/or is used to conduct a reflective review of previously published communications to look for trends and other commonalties.

A simpler tool available for content review is the various readability formulas. The most commonly used ones include the Flesch Reading Easiness (Flesch), the Flesch-Kincaid Reading Level (Flesch-Kincaid Readability Test), and the Gunning Fog Index. All three indexes take a sample of written text, usually at least 100 words long, and analyze it for content and grade-level comprehension. The Flesch and Flesch-Kincaid tools use word length and sentence length as key markers and add different weighting factors. Both were developed by Rudolf Flesch in the 1950s.

The Gunning Fog Index determines the rough number of years of education a person needs in order to comprehend a written sample on first reading and is most often used when large numbers of people with varying educational levels will be reading the communication. This index was developed in the 1950s by Robert Gunning (Gunning Fog Index).

All three of the above indexes require calculations by established formulas of distinct written samples and, similar to content analysis, can be labor intensive. However, numerous software products are readily available to perform the calculations easily and can provide the communicator with a quick analysis to compare to the audience profile. Adjustments can be made prior to message delivery if the readability index does not match the expected audience level.

EVALUATING THE COMMUNICATION EVENT

In the previous section, attempts to benchmark the specific message for readability against standardized measures are discussed as an important method to evaluate message content. Equally important, however, is evaluation of the message event after it has occurred. In general terms, evaluation efforts measure the effectiveness and impact of the communication event and help the organization determine whether or not the goals and objectives of the messages were met. For that reason, every communication event should contain several evaluative measures, both internal and external. Lundgren and McMakin (2004) suggests the importance of the measures will help an organization achieve the following objectives:

- Clarify the level of audience comprehension
- Determine the audience's agreement with the specific recommendations being made
- Identify the audience's willingness to take the recommended corrective action
- Provide documentation for regulatory authorities of compliance
- Understand audience's perceptions of the "helpfulness" of the message

Conversely, poor risk communication efforts may create more problems than they are intended to solve by leading audiences to inaccurate conclusions because they omit key information, emphasize irrelevant information, and produce conflict by eroding the audience's trust in the communicator. Further, they can cause additional and unnecessary alarm or complacency in the audience (Bostrom et al. 1994).

Common measures that would indicate success in the message delivery would include a change (or increase) in the audience's level of knowledge and awareness; support by the audience for the organization's polices or plans; changes in attitudes, opinions, and beliefs; changes in behavior and actions; and an increased level of trust in the organization (Hyer and Covello 2005).

In the mental models approach, described in greater detail in Chapters 3 and 5, the message development process is lengthy, intense, and expensive, but its proponents argue that it provides a much stronger product that succeeds by increasing the level of knowledge about complex and technical subjects in an audience, thereby increasing the probability that the audience will take the recommended corrective action steps (Morgan *et al.* 2002). In an evaluation of a publication created using this approach against a publication prepared by the EPA, both on the topic of radon, concurrent and retrospective evaluations demonstrated that the publications performed better at filling audience knowledge gaps, contradicted misconceptions, and enabled readers to solve problems about radon (Bostrom *et al.* 1994).

The dilemma for most organizations is finding the time and resources to effectively evaluate communication events. The type of validation done on the radon publications is not only costly but also impractical. It is better used in the message creation effort, typically for the types of lengthy written publications intended for wide distribution that work well with the mental models approach.

An additional consideration is the access often required by independent evaluators to inertial documents and information produced by an organization while crafting risk and crisis communication messages. The need to provide unfettered access may occur at a time when the organization is in crisis response mode, leaving the efforts to be, at the least, interfering and at the worst, causing more problems than they are intended to solve. Evaluators may also intrude upon extremely sensitive and confidential information at a time of great risk to the credibility of an organization; therefore, it may be exceedingly difficult for the type of access needed be able to be granted by an organization in the midst of a crisis event.

To increase their validity, evaluation measures should include at least one internal measure and one external measure. Internal measures tend to be far less complicated to utilize and often simply consist of asking a series of questions of the communicators and about the content of the messages. The World Health Organization has published a lengthy document along with a summary field guide to assist organizations with media and other forms of communications during public health emergencies (Hyer and Covello 2005). The concepts, however, apply to other types of risk and crisis communications events. Tables 6.6, 6.7, and 6.8 summarize some of the key internal questions to ask during an internal evaluation.

In addition to the questions noted above, other simple internal evaluative measures include tracking a variety of data sets such as time schedules of employees who participated in the communication events; expenditures for the various communication events, as well as the time needed to prepare for them; work tasks performed and work products available, such as reports and memos; volume of inquiries from the public; number of documents distributed; number of press kits distributed; and number of audience members who participated in the various communication events (Hyer and Covello 2005).

Fortunately, a variety of methods are available to assist in the external evaluation effort that do not intrude upon an organization's functioning level or privacy and are not costly or time consuming (Lundgren and McMakin 2004). Many organizations have made good use of surveys. In the digital communication world, these do not need to

TABLE 6.6. Evaluating Openness and Transparency of Communication[a]

- Were you candid and open with reporters?
- Were you the first to reveal bad news?
- If the answer to a question was unknown or uncertain, did you express willingness to get back to the reporter with a response within an agreed upon deadline (assuming the story was not reported in real time)?
- If you were in doubt did you lean toward sharing more information, not less?
- Were you especially careful when asked to speculate or answer "what if" questions, especially about worst-case scenarios?

Did you:

- Disclose risk information as soon as possible (emphasizing appropriate reservations about reliability)?
- Fill information vacuums?
- Minimize or exaggerate the level of risk? Over-reassure?
- Make corrections quickly if errors were made?
- Discuss data and information uncertainties, strengths, and weaknesses—including those identified by other credible sources?
- Identify worst-case estimates as such, and cite ranges or risk estimates when appropriate?

[a]Hyer and Covello 2005, p. 44

TABLE 6.7 Evaluating Listening[a]

Did you:

- Target the right audience?
- Miss listening to anybody important (stakeholders or partners)?
- Avoid making assumptions about what viewers, listeners, and readers knew, thought, or wanted done about the risk or the situation?
- Identify the target audience and try empathetically to put yourself in its place?
- Acknowledge the validity of people's emotions and fears?
- Respond (in words, gestures, and actions) to emotions that people expressed, such as anxiety, fear, anger, outrage, and helplessness?
- Express genuine empathy when responding to questions about loss?
- Acknowledge, and say that any illness, injury, or death is a tragedy?
- Use media outlets that encourage listening, feedback, participation, and dialogue?
- Recognize that competing agendas, symbolic meanings, and broader social, cultural, economic, or political considerations often complicate the task of effective media communication?
- Display sensitivity to local norms, such as speech and dress?
- Review available data and information on what people were thinking before media interviews?

[a]Hyer and Covello 2005, p. 44

TABLE 6.8. Evaluating Clarity[a]

Did you:

- Speak at the appropriate level of comprehension for your target audience?
- Keep your sentences short and focused?
- Use clear, nontechnical language?
- Use graphics and visual aids to clarify messages?
- Respect the unique media communications needs of special and diverse audiences?
- Translate and test messages to meet the cultural and language needs of special populations?
- Consider how to best stress messages intended to have global reach?
- Personalize risk data?
- Use examples and anecdotes that made data come alive?
- Acknowledge and respond to the distinctions that the public views as important in evaluating risks?
- Use risk comparisons to help put risks in perspective and avoid comparisons that ignored the distinctions people consider important?
- Identify specific actions that people could take to protect themselves and maintain control of the situation at hand?
- Strive for brevity?
- Offer to provide needed additional information within the reporter's deadline?
- Provide the reporter with information about actions that were under way or that could be taken?
- Promise only that which could be delivered and then follow through?

[a]Hyer and Covello 2005, p. 45

be a traditional paper format that is either mailed to the organization or completed at the end of a communication event, although these are effective methods. E-mail surveys are relatively easy to utilize and online surveys that ensure respondents' privacy are easily created on many organization's websites, as well as through numerous free or inexpensive independent websites (e.g., Survey Monkey, Constant Contact). It should be noted that the development of the actual survey questions may be a more intensive process and require the assistance of someone with evaluation expertise; otherwise, the questions may be construed to present the organization in the best possible light rather than answer importance questions about the validity of the communication event. However, even with some limited expertise, a solid evaluation survey can be constructed and can, with similarly limited ongoing assistance, be reused for future communication events.

Telephone surveys are another inexpensive method of message evaluation. As with online or paper surveys, care must be taken to develop the questions asked of respondents and the expertise that might be needed to do so. The cost of using trained surveyors to make the calls increases the expenses of this mode of evaluation. However, as with surveys, agencies exist that provide customized telephone surveys for varying fees that cover the cost of all of the leg work and implementation work of a phone survey process.

For both paper/online surveys and phone calls, additional care must be taken to obtain information from a sample that is valid with respect to the population the communication event was designed to reach. The need to gather demographic data in order to assure that the respondents match the intended audience in key attributes is important. As is noted above, outsourcing some of the work of these efforts may be necessary. In addition, local universities may offer low-cost or free assistance in all phases of the evaluation efforts; students in graduate and doctoral level programs are frequently in need of topics for research studies and/or practicum that offer students who are learning the science of evaluation hands-on experience.

A more intensive approach to evaluations can be found in the use of in-person interviews and focus groups. An organization that chooses to utilize this type of evaluation method will need to be prepared to invest more time and resources in conducting it properly so that the results are accurate and useful. Once again, outsourcing this to an organization with specific expertise may be a sound decision and is readily available in most areas.

REFERENCES

Bostrom, A., C. Atman, B. Fischhoff, and M. Morgan. 1994. "Evaluating Risk Communications: Completing and Correcting Mental Models of Hazardous Processes, Part II." *Risk Analysis* 14(5):789–797.

Coombs, W.T. 1999. *Ongoing Crisis Communications: Planning, Managing, and Responding.* Thousand Oaks, CA: Sage Publications, Inc.

Covello, V. 2008. "Risk Communication: Principles, Tools and Techniques." Posted online at http://www.maqweb.org/techbriefs/tb49riskcomm.shtml on February 25, 2008. Accessed on October 29, 2009.

Fearn-Banks, K. 2007. *Crisis Communications: A Casebook Approach*, 3rd ed. Mahwah, New Jersey: Lawrence Erlbaum Associates.

Flesch-Kincaid Readability Test. Accessed on May 3, 2010 from http://en.wikipedia.org/wiki/Flesch%E2%80%93Kincaid_readability_test.

Gesualdo *et al.* 2010. "Surfing the Web During Pandemic Flu: Availability of World Health Organization Recommendations on Prevention." *BMC Public Health 2010* 10:561. doi:10.1186/1471-2458-10-561.

Gunning Fog Index; Accessed on May 3, 2010 from http://en.wikipedia.org/wiki/Gunning_fog_index.

Hyer, R. and V. Covello. 2005. *Effective Media Communication During Public Health Emergencies.* Geneva, Switzerland: World Health Organization.

Leung, M.L. and A. Nicoll. 2010. "Reflections on Pandemic (H1N1) 2009 and the International Response." *PLoS Med* 7 (10): e1000346. doi:10.1371/journal.pmed.1000346.

Lundgren, R.E. and A.H. McMakin, A.H. 2004. *Risk Communication: A Handbook for Communicating Environmental, Safety, and Health Risks*, 3rd ed. Columbus, OH: Battelle Press.

Morgan, M.G., B. Fishhoff, A. Bostrom, and C.J. Atman. 2002. *Risk Communication: A Mental Models Approach.* New York: Cambridge University Press.

U.S. Department of Health and Human Services. 2006. "Communicating in a Crisis: Risk Communication Guidelines for Public Officials." Washington, D.C.

U.S. Environmental Protection Agency. 2005. "Superfund Community Involvement Handbook." EPA 540-K-05-003.

U.S. Environmental Protection Agency. 2007. "Effective Risk and Crisis Communication uring Water Security Emergencies." EPA 600-R07-027.

7

WORKING WITH THE MEDIA

There exists a unique and sometime thorny relationship between an organization trying to communicate risks to its audiences as it undergoes a full-blown crisis and the media, so much that an entire chapter is devoted to this topic. Even so, it barely scratches the surface of the intricacies of the relationship and the concepts necessary for success in this arena by risk communicators. This chapter will not provide an exhaustive look at media relations but will attempt to address key components that need attention by any organization preparing risk and crisis communication messages.

The ironic reality of media relations is that many organizations struggle to get the media to pay attention to stories about them during normal operations but will find it fairly easy to get the attention of the media in times of crisis or an event that potentially reflects poorly on an organization's functioning. In fact, the media will often be readily available and quick to publish or broadcast a story, whether the organization is ready or not. The key then is to be prepared for those situations when the media call and want a quote for an upcoming new story or even a full-out interview with a high-level executive. The message maps that were discussed in Chapter 5 are one way to be prepared with actual messages for spokespersons; formal written crisis communications plans (addressed later in Chapter 8) provide another internal template for how an organization should respond in a crisis and should include a section on media relations.

Risk and Crisis Communications: Methods and Messages, First Edition. Pamela (Ferrante) Walaski.
© 2011 John Wiley & Sons, Inc. Published 2011 by John Wiley & Sons, Inc.

Preconceived ideas about media outlets and their essential fairness notwithstanding, how the media cover stories and what stories they choose to cover are generally beyond the control of most organizations. There are no simple solutions presented in this chapter to assist an organization in creating a more controlled media environment. What is offered here is a set of guidelines, rules of thumb, and strong suggestions that, when taken as a whole, can improve the overall effectiveness of the response by an organization. As is noted in the following quote, nothing about media relations is an exact science:

> All of these imponderables of reporting and news coverage make communicating with and through the news media an imprecise endeavor. What you say may not be what is determined to be news. How you say it may lead to confused and confusing reports and misinterpretations. Whatever you say is likely to be balanced against opinions that are different than yours (U.S. DHHS 2006, p. 36).

LEVEL OF ORGANIZATIONAL EXPERTISE

In many larger organizations, specially trained individuals fulfill the role of public relations or even more specifically, media relations, and in crises, the public information office. In the event that an organization does not possess this expertise in-house, a higher level executive may take on the role, with participation from many, including the safety, health, and environmental (SH&E) professional.

For those SH&E professionals with in-house media expertise, it is critical that a positive relationship be developed and maintained on an ongoing basis. At the very least, the in-house staff should be well versed in the role of the SH&E staff and how to actively promote the importance of their role and essential activities within the organization. When a crisis occurs, the health and safety of the workforce and/or community are often involved, requiring the knowledge and expertise of the SH&E staff as the messages are being crafted, modified from prepared messages, or delivered. Not having to begin to develop a working relationship at a remedial level with in-house staff at the time of a crisis can be extremely important and increases the effectiveness of the communication efforts during crisis communication events (Hurns and Tapp 2010).

If an organization does not utilize in-house staff expertise, training in media relations is both readily available and highly advisable for several key staff, including spokespersons and other technical support personnel who may be actively involved in the crafting and delivery of messages. Even though many organizations will seek outside consulting expertise in times of crisis, having some internal staff with a working knowledge of how to respond can be beneficial and provide at least some semblance of appropriate response in the early hours and days of a crisis. Their ability to lay the groundwork and implement it during the risk and crisis communications process is invaluable.

A secondary resource that can be called upon for assistance with media relations are trade associations, which, due to their size and purpose, often employ dedicated staff to promote the members of the organization in a positive light and develop ongoing relationships with media representatives that can be utilized when crises occur. Further, many such organizations regularly write articles and press releases about the activities of their individual member organizations, their members as a whole, or the key functions of their industry, thereby creating a ready supply of media information that can be disseminated by the organizations. In addition, the job of the paid staff is to serve the members, so questions about how to handle certain risk situations or assistance with the development of crisis communication plans are among the many activities they are tasked to perform. Finally, trade associations often offer various training programs to their members, either as an open enrollment course at strategic locations or by providing customized training to members at the home sites. These training programs often include media relations.

ADVANCE DEVELOPMENT OF RELATIONSHIPS WITH THE MEDIA

As with any level of crisis planning, developing a working relationship with members of the media in advance of an emergency can serve the purpose of better preparedness as well as create a familiarity with people and organizations. The sound foundation these relationships provide profoundly improves the cooperation necessary in the event of a crisis. The same holds true for media relations; cultivating positive productive relationships with representatives of local media outlets in advance of, and unrelated to, a crisis is strongly recommended. These pre-crisis activities may provide the critical difference in getting your organization's message out in a way that informs the audience but does not devastate the organization (Fearn-Banks 2007; Hurns and Tapp 2010; Hyer and Covello 2005). The value of the relationship is that it improves the balance regarding the needs of both sides: the organization's needs for the media to provide a channel of communication to large numbers of the audience and the media's need to gather accurate and timely information for their stories. When developed in advance, at a time when the pressures of a crisis are not paramount, both needs are met with less frustration, and the relationship is more likely to be built upon trust, respect, and an understanding of the mutual benefit the relationship provides.

Developing relationships in advance typically requires the initiation of the organization. Contacting local media representatives and offering expertise for a story or suggesting a story, even if the suggestion is not taken up, begins the process. Often organizations find that it takes several attempts to both assure the media outlet that it has a sincere interest in working together and also to rise up to the top of a media outlet's many story opportunities that are pursued at any given time. In addition, regular press releases widely distributed at times of positive news about the organization also builds on a familiarity between the organization and the local media. Providing detailed information in press releases also helps with text for a story. All of these actions also promote the organization as one that is open to an investigation into how it operates and is operating at a trustworthy level.

It is also absolutely critical that access by the media to key organizational members be simplified, not just during a crisis but at all times. Assuring that a high-level representative of the organization is readily available 24 hours a day when the media call will further position the organization as open and responsive. A complicated and con-voluted voice mail system that relegates the media representative to leaving a message is frustrating and may lead to a simple hang up, so it is critical that all information distributed about the organization have a key point of contact listed with a phone number that is readily answered. The organization's website should also intuitively lead a reporter to a simple point of contact. A reporter with a deadline in need of supporting information or a quote for the story being written should be able to get to a live person easily.

THE VARIOUS ROLES OF THE MEDIA

It is overly simplistic to state that the role of the media is to report information to a larger audience. Of course, that is a fundamental part of their task, but in current society their roles and responsibilities, and goals and objectives go far beyond that. Regardless of whether one believes that the media have become nothing more than a path for biased information flow and are bent on changing public opinion toward their particular bias, the media in our society have always existed to perform a number of roles; therefore, the roles the media fulfill in risk and crisis communication events are no different. Lundgren suggests their roles include the following (Lundgren and McMakin 2004):

1. **Reporting existing information.** This role is the more traditional one taken on by mainstream media and consists mostly of gathering facts and reporting them out in an interesting way to a particular segment of the audience.

2. **Influencing the way an issue is portrayed.** This role can be less obvious. The number of stories published on a particular issue, the length of each story, and the placement of the story in a newscast or in print all are major factors that can benignly affect an audience's knowledge of an issue and, less obviously, its opinion of the issue. This is true even if the specific media outlet is quite factual in its reporting; the quantity and placement of stories about a specific risk or crisis can have a profound effect on opinion. Crises with national implications are often reported on several times each day by print media, hourly by many broadcast media, and 24 hours a day by other online media and some broadcast sources. This intense level of coverage can last for days, weeks, and even months, depending upon the situation.

3. **Independently bringing an issue to the public's attention or restricting its coverage.** This role is related point 2 above. Many media outlets run feature stories that are independent of any news events that may be occurring. These stories are often termed "human interest" stories; however, there are no real criteria for determining what the public might be interested in other than ideas generated by the reporter and the editorial staff (or, as suggested above, by an organization itself). They often have their genesis in some topic of "human interest," but at times come completely from the particular interests of those in charge of deciding what stories get written, namely editorial staff of the media outlet.

The opposite may also be true in that the media may determine that a particular story does not warrant any mention or will "bury" the coverage in such a way that the audience receives little information about it. In general, this role is restricted to stories that have limited large-scale interest among a wider audience. In short, if most of the major media outlets are reporting on a particular story, restricting its coverage can be more harmful to the outlet not doing so than anything else.

4. **Proposing solutions to a risk-related decision, including taking a stand on an issue.** This role is both active (when a media outlet composes an editorial and makes clear what its specific position on an issue is) as well as passive (when its coverage makes its position fairly obvious). The latter is related to points 2 and 3 above and is observed by the number and content of the stories presented on an issue.

These roles are taken by the media on an ongoing basis and involve reporting on the risks to the audience, which are being conveyed by an organization, as well as reporting on the actual crisis that erupts; the latter is more commonly thought of as "news." As noted in the following sections, an organization's media relations efforts need to encompass both situations as part of an effective risk and crisis communications effort.

Regardless of all of the varying roles the media fulfills, its goals and objectives are fairly simple to categorize. The media believe they know their audience's basic demographics, whether the identified audience is the general public or some specific segment of it. The media also believe they know what their audiences want to hear about and what issues they are concerned about and wish to be informed about. In the search for information for their stories, the media want to collect information that will help them tell their story in a way that affects their audience in some discernable way and gives it the information it wants and believes it needs. This requires a collection of facts and key information that can be turned into a story; the organization becomes the mechanism through which these facts can be channeled. Finally, the media want to create a story their audiences will care about, as well as create emotion and attachment to an issue or an event. When the media's goals and objectives are in line with those of the organization, positive stories are more likely to be published, and vice versa (Hurns and Tapp 2010).

CONSTRAINTS OF THE MEDIA AND MEDIA REPRESENTATIVES

In addition to the roles played by various media representatives, the nature of their jobs places some additional constraints on them. Below is a summary of some of the more common restrictions and typical personality traits (Hyer and Covello 2005):

- **Subject-matter expertise.** Many reporters lack expertise in highly technical subjects. They may do research prior writing a story or conducting an interview, but the time they have to do such preparatory work is limited and the information they gather ahead of time may be incomplete, which might lead them to draw incorrect factual conclusions about a situation.

- **Resources.** Even large media outlets are limited in resources that enable their reporters to prepare in advance. This may also impact their ability to get a reporter to the site of a crisis. This is much more likely to be the case in recent years as the budgets of newsrooms have been cut, severely limiting resources and forcing editorial staff to make very tough choices about which stories to cover and where to send their staff.
- **Career advancement.** The need to move onto larger markets often causes fairly high turnover among reporters. The lack of a stable workforce makes it difficult for most organizations to develop the key relationships discussed above. Smaller markets in particular tend to only be able to attract and maintain staff of limited experience and expertise.
- **Competition.** This exists among reporters for the same media organization as well as between organizations and can be a source of inaccuracy in the rush to get the story out first, as well as sensationalism in an effort to draw the largest viewership. Even major media figures such as Dan Rather and Oprah Winfrey have fallen prey to this problem.
- **Deadlines.** Media deadlines are often extremely tight and unrelenting. As noted above, the rush to meet them can easily compromise the ability to corroborate factual information or do the research necessary to fully understand all aspects of the story. The need to quickly move on to the next story can create a series of short stories that never really follow a situation through to the end but focus mostly on quick stories that highlight major points designed to capture an audience's immediate attention.

WHAT THE MEDIA NEEDS FROM AN ORGANIZATION

It is true that in the search for information to write a story journalists generally need to answer the six basic questions: who, what where, when, how, and why. All of the information they seek from various sources is in an attempt to answer, explain, and elaborate on those key questions, and the use of factual information is vital to this effort. However, interest in a story by the reader is created when contradicting facts can be identified and presented and when emotion or drama can be created in the telling of the story. An organization hoping that the media will publish a story that provides only their set of facts, without presenting opposing views, is certain to be disappointed.

However, what is often not considered is that, while in search of various pieces of information, representatives of the media want something the organization has. This equation places the organization in a position of power and control, which, although slight, can backfire if not used wisely. This is due in part to the information that may be available from other sources with contradictory facts, but also because the media outlets benefit from a healthy degree of curiosity and skepticism when gathering their information. This aspect helps balance the power equation as corroboration of essential factual pieces of information is typically necessary for most media stories; in other words, just because an organization tells the media something is true does not automatically ensure that it will be reported exactly as such unless, and until, another credible source can verify it.

In addition, different types of media have different time needs. In general, print media have more space to fill on any given day, even though the prime space may be at a premium, particularly during a crisis of national proportions. This obviously leads to a need to cover stories in greater detail and to cover additional aspects of the story, even as the crisis unfolds. Despite the increase in the number of television news stations and the constant nature of coverage by some of them, TV news stations remain outnumbered greatly by the number of print media sources. Furthermore, regardless of the national repercussions of the event, competition for time and space on a newscast significantly shortens both the time of the broadcast story as well as the level of detail it can provide. (The latter statement is prefaced by the caveat that there are numerous cable and network news channels whose coverage of any event rivals that of any print media source.)

The media need information from organizations both when a crisis is occurring and when it is not. The former situation poses real issues of time constraints, limiting much coverage to basic information that informs the audience and suggests decisions audience members may need to make for their own immediate safety or shorter-term protection. Organizations that are able to get the basic facts to the media, either because they have access to the broadest channels for that information or because they have developed pre existing relationships, are likely to have a greater influence on what information appears in the early stages of a crisis, when the media may not have as much time to devote to in-depth investigative reporting. In addition, organizations that have taken the time to prepare messages in advance of the crisis are likely to be more successful in getting key messages to the media faster than those who have not. And as noted above, getting the messages to those who need them is critical to the success of any organization's overriding goals and objectives of basic crisis communications.

For some crisis events—typically those with larger implications for larger audiences—after the initial stags of the crisis have passed and the immediacy of the danger ebbs, the media will have more time to devote to detailed reports on the crisis, as well as extenuating factors and issues related to the crisis, thus providing the audience with lengthier stories that inform at a deeper level. Organizations that have established the ability to provide the media with what they need in the early stages of a crisis and whose information proves both factual and vital to the reporting are most likely to be the ones the media will turn to when they need greater detailed information. Again, as noted above, the development of these types of relationships can take time and energy. An organization that can fill the media's need for information during "slow news days" is much more likely to be contacted by the media first when the crisis erupts. Furthermore, any additional information that can be provided to media sources that help construct side stories on a particular topic or crisis may help avoid stories that stray into areas that organization would rather they avoid or might not portray them in a good light (U.S. DHHS 2006; Lundgren and McMakin 2004).

It is important to note that the quality of individual journalists exists on a broad continuum just as it does for any other group of professionals; some are highly professional, ethical and honest, others are not, and there exists every possible combination and degree in between. Hopefully, the profession is successful in weeding out those who posses the worst qualities, but since this is not always true of any other profession, journalists are not an exception.

Finally, it is helpful to remember than as fellow human beings, journalists possess the full range of emotions about any issue they may be covering. Their objectivity is always what has been expected to set them apart from the "average public" who read their stories; however, in the face of a crisis that may have resulted in human tragedy, destruction of the environment, or any other of the other horrific results of a crisis that befall an organization, absolute objectivity and emotionless reporting is an ideal and not always the reality.

FAIR MEDIA COVERAGE

It is not in the realm of this text to provide the author's commentary on the fairness of current media coverage, to defend any particular media source, or to explain what currently exists in typical media coverage. Allegations of bias and slanted reporting are easy to find. Hostility toward even traditionally respected media sources abounds. Entire websites and blogs have been created to allow everyday citizens to refute media outlets, express their frustration, and even make accusations that at times appear incredible.

All of this does not change the basic goals and objectives of an organization's attempts to use the media to deliver risk and crisis communications. Nor do well-thought-out plans and well-meaning attempts guarantee success by the organization. Sandman (2003) takes the position that, over time, most major news stories involving public risks and crises tend to balance out, focusing less on hype and more on reporting facts and providing information necessary for changing perceptions and decision making by the public. Sandman, Hyer, and Covello also remind risk communicators that the public never really panics following receipt of accurate information:

> However, studies indicate that panic is rare, and that most people respond cooperatively and adaptively to natural and man-made disasters. Panic avoidance should never be used as a rationale for false reassurance or for lack of transparency on the part of authorities (Hyer and Covello 2005, p. 14).

An audience may not agree, its members may be angry, and they may be fearful, but they rarely panic. Therefore, an organization's best strategy is to continue to focus communication efforts on providing factual information so that media report them accurately, thereby creating the likelihood of the best outcome for an organization.

Sandman also suggests that risk communicators keep in mind the potential results of messages being repeatedly broadcast on the audience's perception of the media rather than about individual organizations. It may sometimes seem as though the same sound bite or snippet of footage is repeated endlessly or that multiple outlets are all covering the same story but with substantially different slants. The result may be that the audience's trust in the media is eroded, in addition to the erosion of trust in the organization involved in the crisis event. Emotional overload can also result in denial or apathy, neither of which benefits an organization looking for action on the part of an audience.

DEVELOPING A MEDIA COMMUNICATIONS PLAN

As will be discussed in Chapter 8, the development of a crisis communications plan is critical for an organization that wishes to improve its risk and crisis message development and delivery. The elements necessary in such a plan are detailed in this chapter; however, here the focus will be on a smaller type of communications plan called a media communications plan (Hyer and Covello 2005). In many organizations, the plan is an integrated element of a broader crisis communications plan, but depending upon the type of organization and the frequency with which media relations occur, it may also be an expanded plan that stands alone and is cross-walked for reference within an organization's crisis communications plan.

As with any organizational plan, a media communications plan should begin with goal statements and delineate the plan's objectives to be undertaken to achieve these goals. Most goal statements include references to increasing audience trust, informing and educating the audiences, improving awareness, and developing methods for collaboration with other partners in message delivery.

Plans also typically include the following information:

- Staff roles and responsibilities including leadership for varying types of emergencies and authorized spokespersons
- Procedures for information, verification, clearance, and approval prior to delivery to the media
- Procedures for coordinating with other partners
- Policies regarding employee contacts from media
- Media contact lists that are regularly verified
- Exercises and drills for testing media communications
- List of internal and external subject-matter experts
- Preferred communication channels (telephones hotlines, websites, radio announcements, news conferences)
- Location of repository of prepared messages
- Location of useful documents (fact sheets, FAQs, brochures)
- Task checklist for key time frames (generally at 2, 4, 8, 12, 16, 24, and 48 hours)
- Procedures for reviewing and revising the plan.

GETTING THE ACCURATE MESSAGE OUT

Several experts in the field of risk and crisis communications as well as media relations strongly recommend the use of press kits to disseminate key factual information about an organization both before and during a crisis. Press kits serve many purposes; key among them are to provide journalists with ready sources of factual information, not only about the organization, but also about the topic in general. These documents provide an essential and easy way for the media to corroborate facts being presented in a story and can also add filler when needed. It is essential that press kits be provided to all attendees of invited events such as press conferences or public meetings. Lundgren and Hyer offer the following suggestions for the compilation and content of press kits (Lundgren and McMakin 2004; Hyer and Covello 2005):

- **Don't overstuff.** Make sure the contents inside the kit are lean and relevant. If the kit is overstuffed it is more likely that the documents inside will not be read.
- **Fact sheet.** A one-page document that highlights essential facts about an organization, preferably in bullet form. It should be placed in a position that allows it to be immediately viewed and accessed.
- **Relevant articles or reports.** Copies of previously published articles or reports about the organization or the particular event. These can be directly about the organization or published by and about the event or situation in general.
- **FAQs.** These types of documents provide quick and ready answers to the questions most likely to be asked by the journalist or the audience reading the article, and they provide preemptive information that can help elaborate on the event or the organization. They are crucial to help support the organization's position on an issue or method of addressing a particular crisis.
- **Press release.** This is generally intended to recap the reason for the press event. It is important that an organization without in-house public relations staff outsource the writing of this document so that it reflects current standards followed by most journalists and paints the organization's media relations in a professional light. Along with the fact sheet noted above, this document should be placed in a central and easily accessible location.
- **Biographies.** For those who will be speaking at the event or who may also participate or to prepare significant pieces of information that may be delivered by in-house subject-matter experts.
- **Photographs or other visuals.** These can be general shots of the organization and its operations or images captured regarding the particular crisis. It can also include charts, maps, timelines, diagrams, and drawings.
- **Business cards.** These should be of anyone in the organization who is directly speaking at the event or others who may be available to respond to further questions. It is critical that at least one of these key contacts be of someone who is identified as the point of contact for further inquiries and has the ability to respond 24/7, either directly or within a brief period of time. It is also important to be certain that the phone numbers on all of the business cards identify a line

that answers with a live person after business hours. As noted above, the media's time frames for publication often require quick responses. In the absence of them, the message delivered in the eventual published story may not be at all what the organization wants.

CHOOSING A SPOKESPERSON

The face in front of the television camera and the person who answers the reporter's questions during a one-on-one interview must be chosen carefully. In most cases, the highest-ranking executive is the most logical choice as he/she is the recognized leader of the organization, as well as the one person who has the credibility and authority to speak for the organization and indicate what actions the organization will or will not take regarding a potential risk or a current crisis. It is also best if the organization chooses one spokesperson or at least one primary spokesperson, so that the messages being delivered remain consistent.

Those criteria represent ideal situations and assume that the highest-ranking executive is a capable public speaker and is able to effectively deliver key prepared messages. However, not everyone in a position of authority in an organization is an effective public speaker. They may have difficulty in delivering prepared messages without appearing as though they are reading from a script. They may ramble or not be able to speak in complete sentences, interjecting their comments regularly and annoyingly with "um" and "ah" and "well." They may freeze in front of a camera, look guilty or get defensive under questioning by a skilled reporter who is looking for a headline. Some executives simply do not have a pleasant speaking voice or one in which an audience would respond to as trustworthy and credible. And some simply do not have a pleasant appearance in front of a camera. See Table 7.1 for the necessary characteristics for the person designated as an organization's spokesperson.

These identified criteria also assume that the event is not so significant that it requires multiple daily communication events on a national or international scale, hourly press conferences, news briefings, and speaking opportunities, as well as interviews with various media sources. They also assume that the event is not so long-standing that the ability of one spokesperson to respond and be available for all of the communication events required is impaired; there are ample opportunities to sleep, eat, and work with other members of the organization's executive staff dealing with the crisis event. During the recent BP Deepwater Horizon explosion and subsequent oil spill in 2010, primary spokesperson Tony Hayward made several key errors in his public remarks that many believe led to his eventual removal from his role as spokesman and eventually his actual executive position with the BP organization. However, BP Chairman Carl-Henric Svanberg noted accurately that the pressure of his situation was probably more than most people could withstand without making key errors: "It is clear that Tony (Hayward) has made remarks that have upset people. But he is also a man who has probably been on 100 hours of TV time, maybe 500" (Henry 2010). (For a more in-depth analysis of the crisis communication messages of this particular crisis, see the case study in Chapter 10.)

TABLE 7.1. Personal and Professional Characteristics of a Designated Lead Spokesperson[a]

The designated lead spokesperson should:

- Possess excellent media skills
- Have sufficient authority or expertise to be accepted as speaking on behalf of the organization
- Possess or work to develop good professional relationships with important members of the media and other important partners and stakeholders
- Be:
  - perceived as authoritative and credible by stakeholders, partners, and the public
  - at ease with the media
  - knowledgeable (generally and specifically) about the emergency, its dynamics, and its managements
  - a subject-matter expert on the event or able to delegate to subject-matter experts
  - resourceful
- Be able to:
  - learn quickly
  - respond to sensitive questions within their areas of expertise in a professional and sensitive manner
  - effectively respond to hostile questions
  - stay on message yet remain flexible and able to make decisions quickly
  - offer examples, anecdotes, and stories
  - provide effective on-the-spot responses to media inquiries
  - express technical knowledge or complex information in a way that can be easily understood by reporters and by the average person
  - remain calm and composed at all times
  - express caring, listening, empathy, and compassion
  - work well under pressure or high emotional strain
  - accept constructive feedback
  - share the spotlight
  - call on the expertise of others
  - give thanks to others and distribute praise
  - take responsibility for things that go wrong
  - present the appropriate tone for the audience
  - defer, delegate, and redirect questions to others as needed

[a]Hyer and Covello 2005, p. 21

Finally, all of the above criteria for selecting a good spokesperson assume that the event does not require the delivery of technical information better communicated by the individuals who have the knowledge, experience, and even the academic degrees, that permit them to be viewed as more competent authorities on certain aspects of the event or the organization's response to it. In these types of situations, it may be that a team of spokespeople is best suited to address the event, whether as group or in varying combinations based upon the specific communication event, and that one person is identified as the lead spokesperson or at least appears when others are speaking on behalf of the organization.

The additional members of the organization's team may participate in communication events in a variety of capacities and generally are responsible for providing support and assistance to the subject-matter experts. In incidents of long-standing duration and intensity, a team is critical if the organization expects to keep up with the ongoing demands for information from the media (Hyer and Covello 2005). Other internal members of the media team often include other key executive staff, members of the human resources department, internal subject-matter experts, and SH&E department staff.

Larger organizations often utilize the services of a public information officer (PIO). PIOs are, on occasion, the spokespeople for organizations in many communication events but more often act as the linkage between the organization and the media by providing general information, coordinating requests for interviews, and managing the other members of the media relations team. Strong PIOs fulfill a number of key roles during a crisis and must be successful at the following tasks (Hyer and Covello 2005):

• Demonstrating skills in team building, negotiation, and conflict resolution
• Developing and implementing the organization's media communications plan
• Selecting and prioritizing media outlets
• Compiling media contact lists and expert contact lists
• Training organizational spokespeople (or arranging to have them trained)
• Developing and delivering crisis-specific information to the media, partner organizations, agency or staff employees, and the public

PREPARING FOR AN INTERVIEW

It goes without saying that no one within an organization should attempt to be interviewed by a representative of the media or speak at a communication event such as a press conference without some preparation, the amount and type of which is determined by the details of the situation. In some situations, the organization may have a public relations department or representative who can handle the preparations; in other cases, this expertise may need to be outsourced. Regardless of the individual organization's situation, no one should treat a media interview or communication event as a "normal" conversation (U.S. DHHS 2006).

At the very least the interviewee should be made aware of why the specific media representative called, what are the deadlines for both the interview and the filing of the story, what will be the format of the interview (location, number of people present, on camera or off, etc.), how long the interview is expected to last, what the person's role will be, and what knowledge the reporter may already have on the topic (Hurns and Tapp 2010; U.S. DHHS 2006). Once this information is gathered, the interviewee should locate and review or be briefed on any prepared key messages on the specific

TABLE 7.2. Preparing for an Interview[a]

Topic-Related Questions to Ask	Procedural-Related Questions to Ask
What is the subject or topic of the interview?	Who will be conducting the interview?
What specific subjects does the reporter expect to cover in the interview?	Does the reporter specialize in any particular area?
Would the reporter like suggestions on who else to interview?	When will the story be published or broadcast?
What types of questions will be asked?	How long has the reporter been a journalist?
What specific questions will be asked?	What is the reporter's deadline for the story?
Has the reporter done any background research related to the topic of the interview?	Where will the interview take place?
Who else has the reporter interviewed?	Will the interview be audio recorded or videotaped by the reporter?
How will the reporter use the interview material?	Where will the story appear?
	How long will the story be?

[a]Hyer and Covello 2005

crisis event and have the opportunity to discuss how the messages can be bridged into the likely questions of the reporter. It is also helpful to have clarity on previous media coverage of the event, both in the past and as part of the current situation, to understand the perspective being presented to the audience, as well as any slant the media may currently have toward or against the organization (Hurns and Tapp 2010; Fearn-Banks 2007; U.S. DHHS 2006). Table 7.2 includes tips on preparing for an interview; Table 7.3 includes "trick" questions that may be asked by a reporter.

Finally, opportunities to practice answering expected questions is vital, particularly if the interviewee's experience in media events is limited, when the event is controversial, or when the organization is under fire for its role or response to the event. Standing before a mirror is one simple and practical method that can be performed with limited time and resources. More complicated practice sessions involve lengthy preparatory meetings with a public relations staff person or even conducting full-blown mock press conferences or interviews. Additional discussions should take place regarding the type of clothing to wear; in general, the recommendation is to wear what is appropriate for the position held in the organization. Looking professional doesn't always require a suit, but it does require clean, not rumpled, clothing and combed hair (Hurns and Tapp 2010). Table 7.4 highlights some interview do's and don'ts.

AFTER THE INTERVIEW

Most interviews do not represent the last time a reporter is in contact with an organization. Oftentimes the need to clarify something that was said or ask additional questions may occur while the story is being written. The back-and-forth process can go on for some time depending upon the reporter's deadline, but it is always in the organization's best interest to remain open and available for future contact. Attempts to provide

TABLE 7.3. Typical Trick Questions Asked by the Media[a]

Question	Problems Associated With It
Speculative questions Example: "If the fire had not been put out, how many acres of land would have been destroyed?"	Are dangerous to answer. The simplest answer is to say you are not able to speculate.
Leading questions Example: "You do agree that your company should have known this was going to occur?"	Reporter usually already has the answer and is looking for confirmations. Agree only if you are certain.
Loaded questions that are an attempt to get you to admit to wrongdoing Example: "Isn't it true that you knew your produce was contaminated before it left the manufacturing facility?"	Reporter is often looking for an emotional response. Effective responses are to rephrase the question and answer it. "Are you asking if we knew the product was contaminated? No, we were not."
Uninformed questions that indicate the reporter does not have much information about your organization or the situation Example: "What does your company do?"	Reporter may not have done his or her homework. Provide a press kit with simple factual information.
False questions that contain deliberately inaccurate information Example: "You have lost $20 million as a result of this crisis?" Answer: "No, only $10 million."	These are usually an attempt to get you to divulge information you might not otherwise offer.
Know-it-all questions Example: "I have my story already written. I just need you to verify a few facts."	Reporter may have a viewpoint and is only looking for confirmation or may be trying to get you to release "dirt."
Use of silence	Most interviewees have a hard time with silence and will tend to make their answer longer than necessary or add after they have finished.
Accusatory questions Example: "Isn't it true that your suppliers knew all along that this problem would occur?"	These are usually an attempt to get you to blame others.
Multiple-part questions	These are confusing to you and the public. Break the question apart and answer it separately.
Jargonistic questions	The use of lots of technical jargon or terms is confusing to audiences. Answer the question in everyday language.
Chummy questions Example: "Off the record, do you think . . . ?"	Nothing is off the record, and the reporter is not your friend. This is a professional relationship.
Labeling questions Example: "Would you call this situation devastating for the community?"	The aim is make an issue simplistic or negative. Don't accept the reporter's labels unless they are accurate or you agree.
Good-bye questions Example: "By the way . . ."	Often posed at the end of the interview and may have the appearance of being off the record.

[a]Fearn-Banks 2007, pp. 29–31

115

TABLE 7.4. Interview Do's and Don'ts[a,b]

Be patient, open, and honest	A reporter whose knowledge of a specific subject matter is limited may need to ask several similar questions before clarity is achieved. The issues of openness and honesty are addressed in greater detail in Chapters 3 and 5.
Listen carefully to the question and think before you speak	If you don't understand the question, say so and ask for clarification. Never assume you know what the reporter is asking and don't begin your answer until the reporter stops speaking. A few seconds of silence to gather your thoughts will help ensure a clearer and more accurate answer.
Don't feel obligated to answer every question	Be careful not to stray into areas that are beyond your expertise or authority. If you are not able to answer the question for those reasons, say so and refer the reporter to someone in your organization who can.
If you begin to feel uncomfortable with the directions of the interview, return to your key messages	The reporter may be "fishing" for a scoop, looking for controversy to improve the story, or may simply be asking many questions to gather a lot of information. Try to stay in control of the interview by using your key messages as bridges; always look for ways to pull your answers back to them. If it is clear you will not stray from key messages, a reporter is more likely to give up confrontational questioning.
Avoid opinions or speculation	Factual answers are always the safest and more likely in line with the message the organization wishes to communicate; your opinions are not what the story is about. Speculation is best left to experienced spokespersons or high-level executives.
Use common language	The audience for just about any story written by mainstream media will be the general public and not fellow experts in the area. Common vocabulary is better suited for answers to questions for a story that will be read by the general public. Using highly technical words, jargon, or phrases can confuse the reporter, make it appear as though you are trying to impress the reporter, and may result in a story that is less accurate or more confusing because the basic information was not communicated.
Don't speak disparagingly of others	An interview is not the time to be blaming other organizations, your competition, or regulatory bodies. The media also not the place to fight battles. Stick with the adage "If you can't say something nice about someone, don't say anything at all."
Be prepared to support your facts	If you make a statement, be sure that you can back it up with ready references or printed out documents. It will not only improve your credibility with the reporter, it will make it much easier for the reporter to verify your position while writing the story, increasing the chances that your point of view appears in the final story.
Know what controversial information has already been published	Do your homework on previous coverage of the issue or event and know what controversial topics have been covered, particularly if the coverage contradicts your organization's position. Be ready to refute the information if possible with facts that can be supported and references that can be given to the reporter.

TABLE 7.4. (*Continued*)

Don't take anything personally	Neither the questions asked, the tenor of the reporter, nor the actual published story is about you. It's about the issue or event, even if it appears that your reputation is being challenged.
Don't embarrass the reporter or argue with them	This interview is not a battle of wills; it's about getting your organization's position into the public domain. If you adopt a defensive posture, the interview will likely deteriorate. Likewise, embarrassing the reporter will more likely end up in a story that doesn't serve your organization well. Think of yourself as the guest of the reporter and the whole process will be more likely to flow smoothly.
You are never off the record	As noted above, you are the guest of the reporter, not a friend. Keep the discussions professional and never say anything that you do not want to have in print.
The reporter gets to pick the topic and questions	Accept the role of the reporter in "running the show." When you try to steer the questions or topics into one particular direction, you are likely to run into resistance and pushback from the reporter.
Know what topics are off limits for you	In preparing for the interview, be sure to take some time determining what you cannot talk about as well as what you can. If the questions veer into those areas, politely decline to answer the question, indicating that you are not authorized to speak about it. But don't end the discussion there; offer to find the person in the organization who can answer the question and make sure it happens within a reasonable period of time.
Never say "no comment"	Even if you cannot answer the specific question, avoid those two words as they often imply guilt or a problem that will only spur the reporter on to try and find out what you are hiding. As above, if you cannot answer because you don't know or are not permitted to do so, offer to find the person who can and make sure it happens. If you have properly prepared for the interview, the questions will not come as a surprise. Before you sit down have a plan for how to address them.
If you make a mistake, ask for the chance to clarify	Everyone misspeaks sometimes and/or says something that is incorrect or appears to be. If you do so, stop and ask for the opportunity to clarify. There's no need to explain why you erred in what you said, just be certain to get the corrected version out into the interview.
Check your quotes	Make sure you know in advance whether or not the interview will be recorded and, if it isn't going to be, ask to have any quotes read back for clarity. As above, if the quotes need to be restated, ask for the opportunity to do so. If the quote is not an accurate rendition of what was said, ask for the chance to restate it.

[a]A number of references are available to provide general rules to follow when an organizational representative is interviewed.

[b]References for this table include Donovan and Covello 1989; Fearn-Banks 2007; Hurns and Tapp 2010; Hyer and Covello 2005; Lundgren and McMakin 2004; U.S. DHHS 2006.

answers or sources that can be cited to support the organization's claims are essential. The reporter may be answering to an editor who is being particular or simply wants to be able to support anything that ends up published (Donovan and Covello 1989).

As noted above, reporters are often on tight deadlines, particularly for radio or television stories. Availability means more than simply being available to accurately answer questions or provide sources but also in doing so in a timely manner. Lack of response can be indicative of any number of problems, including an inability to support a claim, confusion within the organization about what the media can know, or lack of organizational structure, none of which are helpful portrayals. The simple reality is that the story will be published, with or without the additional information (Lundgren and McMakin 2004).

Finally, after the story is published, make sure to take time immediately to review it for accuracy or completeness. If it is a broadcast story, tape it so that it can be reviewed. If any errors are noted either in what the spokesperson said or in how the story was reported, make sure a timely call to the reporter or their editor occurs so that corrections can be made. If the error is egregious, additional action might be necessary, including seeking legal counsel or asking for a formal correction (Donovan and Covello 1989). Table 7.5 details steps to take when correcting media mistakes.

TABLE 7.5. Correcting Errors in Media Reporting[a]

If there is an error in the story:

- Remain calm and composed when speaking to reporters or editors about errors and mistakes.
- Contact the reporter directly and point out errors only if the error is significant.
- Do not complain about trivial mistakes or omissions.
- Ask the reporter to amend the office file copy of the story.
- Consider asking the reporter to make an appropriate change in their next story. (Note, however, that this can be controversial and lead to a difficult relationship with the journalist.)
- Avoid embarrassing the reporter by naming them during a news or press conference or briefing.
- Avoid if possible going to the reporter's editor or producer—this should only be done if there is a major mistake and if the reporter will not acknowledge the mistake and make the requested correction. By going over the reporter's head, you may ruin any working relationship you have developed.
- If the error occurs in the stories of several different reporters, or if the story is picked up by a wire service, and if the error is deemed major, then correct the error during the next news release, media briefing, or news conference without naming the individuals responsible for the error.
- Recognize the difference between errors and differences in points of view—differences in points of view will generally not be corrected.

[a]Hyer and Covello 2005, p. 40

REFERENCES

Donovan, E. and V. Covello. 1989. *Risk Communication Student Manual*. Washington, DC: Chemical Manufacturers' Association.

Fearn-Banks, K. 2007. *Crisis Communications: A Casebook Approach*, 3rd ed. Mahwah, New Jersey: Lawrence Erlbaum Associates.

Henry, R. 2010. "Questions About Who's in Charge of BP Oil Response." Associated Press, June 19.

Hurns, D. and L. Tapp, L. 2010. "Working With the Media: Telling Your Story Effectively." *Professional Safety* 55(1):52–54.

Hyer, R. and V. Covello. 2005. *Effective Media Communication During Public Health Emergencies*. Geneva, Switzerland: World Health Organization.

Lundgren, R.E. and A.H. McMakin, A.H. 2004. *Risk Communication: A Handbook for Communicating Environmental, Safety, and Health Risks*, 3rd ed. Columbus, OH: Battelle Press.

Sandman, P. 2003. "Fear Factory: Have the Media Overblown Canada's Health Scare?" Posted online at http://www.petersandman.com/articles/Toronto.htm on April 11, 2003. Accessed on May 3, 2008.

U.S. Department of Health and Human Services. 2006. "Communicating in a Crisis: Risk Communication Guidelines for Public Officials." Washington, D.C.

8

DEVELOPING A RISK AND CRISIS COMMUNICATIONS PLAN

In order to function effectively during a crisis event, organizations must have a written plan that delineates the roles, responsibilities, key players, and procedures. The planning process and its eventual document will vary widely from one organization to another, as it needs to be based upon an internal needs assessment, the size of the organization, the severity and probability of likely crises, and other factors. The planning process traditionally begins with some type of formal, and preferably quantitative, risk assessment from which the written document is derived. This chapter will begin by providing a brief summary of the more common tools and techniques available for performing risk assessments and then delve more deeply into the content of a crisis communications plan (referred to hereafter as the "Plan"), including a discussion of planning guidelines, planning process suggestions, and key plan elements.

It should be noted that the Plan, as it is described in this chapter, is intended to focus solely on the communication process an organization undertakes when a crisis occurs. It does not detail the actual response of the organization to the crisis; that is more appropriately handled in a comprehensive crisis management plan. The Plan discussed here either functions as an element of the comprehensive crisis management plan or as a stand-alone document that is part of an organization's overall crisis response.

Risk and Crisis Communications: Methods and Messages, First Edition. Pamela (Ferrante) Walaski.
© 2011 John Wiley & Sons, Inc. Published 2011 by John Wiley & Sons, Inc.

DEFINING ACCEPTABLE RISK

Prior to determining the proper tool for the organization's risk assessment process, consideration must first be given to how the idea of "acceptable risk" will be approached. This term and what flows from it has recently received substantial attention from the safety, health, and environmental (SH&E) professional community (Manuele 2008, 2009, 2010; Hansen 2008). In his writings on the subject, Manuele documents numerous standards and guidelines utilized worldwide that make frequent use of the term, yet his experience is that SH&E professionals are wary of using the term. He believes this occurs because of the limits of understanding the nature of risk, concern over subjective judgments made when risks are assessed, lack of statistical probabilities that allow risks to be numerically categorized, and lack of experience in significantly hazardous environments where "nontrivial" risks are accepted as part of daily operations.

For purposes of this chapter, the term "acceptable risk" is defined as: "That risk for which the probability of a hazard-related incident or exposure occurring and the severity of harm or damage that may result are as low as reasonably practicable and tolerable in the setting being considered" (Manuele 2009).

Hansen agrees and adds that acceptable risk will vary among and within organizations and must be evaluated within the context of what the audience views as acceptable. Further complicating this idea is the reality of opinions and attitudes: "Because attitudes about the acceptability of risks are not consistent, there are no universal norms for risk acceptability. What your stakeholders view as an acceptable risk will depend upon a number of factors" (Hansen 2008, p. 50).

Those factors include the following:

- Nature of the risk
- Who or what is at risk
- To what degree can the risk be controlled
- What is the risk taker's past experience

It is important to note that all four of those factors correspond closely to several of the 15 risk perception factors noted by Covello in the risk perception model described in Chapter 3. Figure 8.1 addresses the concept of acceptable risk to an audience.

RISK ASSESSMENT TOOLS SUMMARY

It is not the intention of this chapter to provide a comprehensive discussion of how to choose the appropriate risk assessment tool and then detail the process for its application, assessment, and implementation. The topic of risk assessment and the tools utilized in the process is extensive, and the reader is referred to numerous texts specific to the topic (Manuele 2008; Haight 2008; Covello 1993). Several of the risk assessment tools require fairly extensive expertise in order to properly apply and implement them, which may often be beyond the ability of internal organizational staff. In such cases, outsourc-

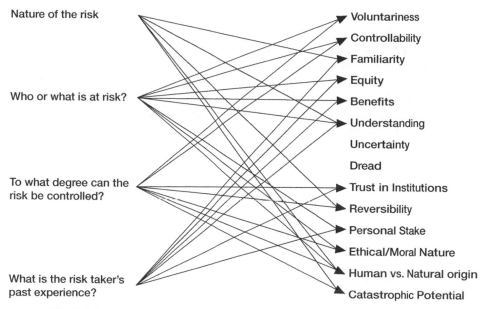

Fig. 8.1. Risk Assessment—The Concept of Acceptable Risk to the Audience

ing the risk assessment process may be the best alternative, depending upon the tool selected. In addition, the purpose of risk assessment in the development of a written risk and crisis communications plan is limited to helping the planning team understand what risks and crises are being addressed in the Plan in a fairly general manner.

The list below provides a brief summary of three common risk assessment tools that can be utilized by the team crafting the written Plan (Hansen 2008; Manuele 2008). For a more detailed discussion of additional tools, the reader is referred to the sources quoted in this chapter.

1. **Checklists.** A checklist represents the simplest and most cost-effective method of risk assessment. It can be developed internally and requires little training to complete. The quality of the checklist developed is the variable that determines the effectiveness of its use and its eventual applicability, so a reasonable period of upfront development time is often necessary unless a prepared checklist is already available. The checklist format walks the organization through a set of defined criteria typically derived from published standards, codes, or industry best practices. The checklist is often in a question format with the appropriate answers being "yes," "no," or "not applicable," with a comments section for additional discussion.

   Once completed, a checklist offers the organization a summary of the risks that exist, albeit with limited discussion about their severity and probability, except for what might appear in the comments section. Its value is the ease of development and the limited need for expertise in the method. The downside is that it takes some time upfront to develop and provides limited quantification of the risks.

2. **"What-if" checklist.** The strength of the checklist tool can be enhanced by combining it with a "what-if" process, the latter of which uses brainstorming sessions among organizational representatives to identify potential hazards, which are then transcribed to a checklist format. (It should also be noted that the "what-if" tool can be used independent of the checklist tool.) The use of a skilled facilitator helps this tool to be more comprehensive by helping the workgroup delve deeper into the potential hazards by asking "what if" repeatedly until the topic has exhausted possible hazards.

   As with the use of a checklist technique, this is a tool that can be utilized with existing internal resources and limited formal training beyond helping the team members understand how to brainstorm.

3. **Matrices.** A matrix provides a method for analyzing a risk based upon two variables, typically severity and probability. It is a fairly easy tool to utilize internally with minimum expertise. The definitions for the variable vary greatly and can be customized to the organization's preference. Completion of a matrix simply requires the presence of those in the organization tasked with assessing risk. A facilitator can be chosen to help the group brainstorm about the various risks that may occur and the placement of them within the matrix. Once completed, the organization determines which of the risk categories it will focus on in terms of plan development and/or mitigation strategies. Typically not all risks are planned for; those with both low probability and severity may not provide a benefit in terms of time and monetary expenditures.

In Fig. 8.2, the variables are simply defined by terms "low," "medium," and "high." An organization would need to clarify internally how those terms applied and the use of a simpler matrix implies a greater degree of subjectivity on the part of those completing it.

In Fig. 8.3, the variables are defined in greater detail and an attempt is made to assign a numerical rating to the scale in order to increase the objectivity.

For a comprehensive treatment of the matrices approaches to risk assessment the reader is referred to the standard published by the U.S. Department of Defense (2002). Table 8.1 compares common risk assessment methods.

KEY PLANNING GUIDELINES AND PROCESSES

Before beginning the process of writing the Plan, an organization needs to ensure that it is prepared to implement several key planning guidelines and has a solid understanding of the purpose of the Plan. According to a planning document prepared and published by the Centers for Disease Control: "The plan is not a step-by-step 'how-to.' It is the bones of your work. It should systematically address all of the roles, lines of responsibility, and resources you are sure to encounter as you provide information to the public, media, and partners during a public health emergency. More than anything,

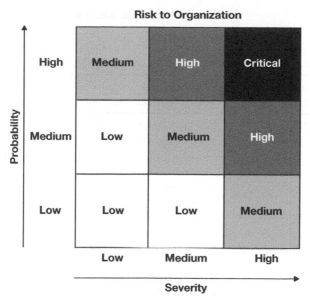

Risk to Organization

Probability ↑			
High	Medium	High	Critical
Medium	Low	Medium	High
Low	Low	Low	Medium
	Low	Medium	High
	Severity →		

Fig. 8.2. Simple Risk Matrix

PROBABILITY

	Insignificant Injury	Minor Injury	Minor Injury	Illness - Injury	Illness - Injury	Major Injury	Major Injury	Single Fatality	Fatality	Multiple Fatalities
Certain	10	20	30	40	50	60	70	80	90	100
Almost Certain	9	18	27	36	45	54	63	72	81	90
Very Likely	8	16	24	32	40	48	56	64	72	80
Probable	7	14	21	28	35	42	49	56	63	70
Likely	6	12	18	24	30	36	42	48	54	60
Likely	5	10	15	20	25	30	35	40	45	50
May Happen	4	8	12	16	20	24	28	32	36	40
Improbable	3	6	9	12	15	18	21	24	27	30
Unlikely	2	4	6	8	10	12	14	16	18	20
Very Unlikely	1	2	3	4	5	6	7	8	9	10

SEVERITY

KEY			
Not Significant		0-3	May be ignored, no further action required
Very Low		4-12	
Low		13-25	Ensure safe working
Moderate		26-42	Refer to risk assesment, safe working procedures
High		43-67	Monitor control measures
Very High		68-100	Avoid if possible, full method statement if not

Fig. 8.3. Risk Assessment Matrix with Weighted Values

TABLE 8.1. Comparison of Common Risk Assessment Methods

Method Type	Expertise Required to Apply	Benefits of Method/Results	Detriments of Results
Checklist	Minimal	• Can be completed someone with limited expertise • Provides a simple listing and brief understanding the risks a company is likely to encounter	• Quality of checklist determines quality of result • Provides limited analysis of the individual risk and no means to compare risks • Does not require input from various members of the organization
"What-if" checklist	Minimal	• Expands upon simple checklist and evaluates the various risks by imagining what would happen to the organization if they occurred • Can be completed within the expertise of the organization • Requires input from multiple organization members	• Quality of checklist determines quality of result • Provides limited analysis of the individual risk and no means to compare risks
Matrix	Minimal	• Can be completed within the expertise of the organization • Provides a simplified means of evaluating risks on the critical variables of severity and probability • Levels of quantification can be expanded by the team completing the matrix • Can be used to compare and contrast various risks	• Does not require input from various members of the organization • Comparison of risks is limited to review of two variables

your crisis communications plan is your resources of information—the 'go-to' place for must-have information" (Reynolds 2002, p. 69).

While each organization's written planning documents for emergency and crisis responses will vary, the Plan will generally fit into the overall framework in one of two ways: either it exists as a stand-alone document and is cross-referenced to other emergency plans, or is a separate element within an organization's overall crisis management plan. Both formats work and neither has any particular benefits. The content is what defines and differentiates a high-quality Plan, as does the accessibility of the Plan by those who need it when they need it.

As noted above, the Plan needs to begin with some type of risk assessment that defines what types of risks and crises the Plan will cover. This information belongs at the very beginning of the Plan and will be addressed in greater detail later in the section that discusses plan elements. The Plan also needs to have management support in place before the development process begins. This support takes several forms, the first of which is a written and verbal commitment from the highest executive in the organization about the importance of the Plan and the process to develop it. Management support also takes the form of dedicating staff time to develop the Plan and ensuring there are financial resources available to implement it. This commitment will vary among different organizations but is essential before moving forward. Any planning process that begins without it will most assuredly be wasted time and will produce a document that is never used except when needed to document simply that it exists.

As noted above by Reynolds, crisis communications plans are intended to be utilized during a crisis, which necessitates that they are dynamic and minimalist documents that provide general direction. The document must be the authority for crisis team members, provide a summary of important information, and guide their activities during the crisis event, rather than a detailed approach to crisis management. Since each crisis is different and progresses along paths that may not be easily controlled, providing an outline for the team to follow serves its purposes better than a set of length and detailed policies and procedures.

In addition to being a dynamic document during its use, the Plan must also be developed with the understanding that it will need ongoing review and revision if it continues to be a helpful tool for the crisis team. Regular reviews that are planned for both in terms of timing and resources are critical and should occur at timeframes of no less than annually. Other situations that would necessitate a review would be following a crisis that is part of a "lessons learned" review and any major change in the organizational structure or operations. In order to ensure that the organizational changes are accounted for in the Plan, it is a good idea to assign a staff person the task of a regular paper review, perhaps as often as monthly. This regular review results in notifying others only when changes are necessary and convenes the entire planning team only when needed.

It is also critical to note that the Plan should be a controlled document, with additions and revisions made through a formally documented process that allows for tracking and a means to ensure that only the most recent versions are available for general review. Internal staff must be informed that making and distributing copies of the document is limited to one identified staff person or department. Distribution outside of the organization also needs to occur only through channels that are delineated in the Plan. This is not to suggest that the Plan is a confidential, top-secret document, but because of the nature of the activities detailed within it, ensuring that only the most recent document is distributed within and outside of the organization serves to reduce confusion when the Plan needs to be implemented.

Once the management support and direction has been established, a planning team needs to be gathered and tasked. Again, each organization brings a variety of representatives to the process and will vary in the number and expertise of the team members. At the very least, team members should represent key departments such as executive management, human resources, SH&E, and public relations. Other internal members should come from production/operations and internal emergency response teams, if appropriate. Depending upon the level of internal expertise, outside consultants with crisis planning expertise may be called upon to either be planning team members or perhaps even lead the effort. In the latter scenario, it is critical that the external consultants be able to "hand off" the document to operational staff once it is completed, unless the organization expects the external consultants to be a part of the response. If not, the internal staff needs to be able to implement the Plan, literally at a moment's notice.

Lastly, before the Plan can be drafted, more than just management support and risk assessment information needs to be available to the team. Data gathering can be a team task or can be done prior to the first meeting but at the very least team will need to know some of the following:

- Designated staff members (line and management) by title and department of who will be responsible for the actual response.
- Other emergency and crisis plans that exist within the organization and how this Plan will fit into the existing structure.
- What processes are already in place within the organization to determine which information can be disseminated, by whom, and to whom.
- The names of those members of the organization permitted to perform the role of spokesperson and public information officer.
- Representatives from other emergency response organizations who are expected to be part of the Plan team.
- A copy of existing mutual aid agreements (MAE) or memorandums of understanding (MOU) with other local emergency response organizations.
- Any other document, policies, or programs that currently exist that will need to be reviewed and incorporated into the Plan. These would include any prepared key messages. If none are available, the team needs to know if they will be responsible for developing them.
- A list of the audiences expected to be impacted by the implementation of the Plan and existing lines of communication with each audience.
- The resources and facilities that will be available to the team during the implementation of the Plan and the means of securing them.
- Information regarding any pre-training that might be helpful to the Plan team members so that they may function more effectively in their roles.

Not all of the details for each of the points above may be known at the time the team begins to develop the Pan. Some will become part of the information gathering process.

KEY PLAN ELEMENTS

Prior to the actual work of developing the Plan, the goals for the process and product should be established. As noted above, these goals ought to originate within the management of the organization so that the final Plan can be approved and continue to receive the support needed. Most Plans have at least two major goals: to anticipate and prevent crisis when possible, and to respond to crises effectively by providing timely, accurate, and helpful information to identified publics (Reynolds 2002). Beyond these major goals, individual needs of the organization can be addressed.

Numerous formats exist for the outline of the Plan (Fearn-Banks 2007; Lundgren and McMakin 2004; Reynolds 2002). A review of several formats reveals many common elements and few differences. Each organization should undertake a review of several formats before deciding on the one that works for its needs. In addition, any of the formats can be modified to meet unique planning needs or situations; using them as boilerplates helps ensure that key information and content is not missed.

The following Plan outline represents a culling of several formats and presents the best overall outline that will address the needs of most organizations (Reynolds 2002; Lundgren and McMakin 2004; Fearn-Banks 2007):

- **Title page.** This page should include some reference to the initial approval date of the Plan as well as the current version and its associated approval date. This page may also include a summary of all Plan reviews/revisions and their associated dates or this information may be listed elsewhere in the Plan. As noted above, the organization should also maintain a log of the various revisions, noting the changes in summary form. This log does not necessarily need to be part of the Plan but should be referenced somewhere and should be readily available for review. Finally, the signature of the key executive approving the Plan should appear at the bottom of the title page. Alternatively, a good way to demonstrate the organization's support for the Plan is to follow the title page with a brief statement from the key executive detailing the importance of the Plan and its association with the organization's overall mission statement, followed by a signature and date.
- **Authority.** This section is a comprehensive list of all regulatory and statutory requirements that affect the organization and require the development of the Plan.
- **Purpose.** This section answers the question of what the Plan is intended to do, for, and by whom. This section should also include a brief statement of the management support of the Plan and how it fits into the organization's overall mission (if not addressed on a separate page as noted above). It can also detail the philosophical position the organization takes with regard to crisis communications.
- **Scope.** This section provides a list and/or description of what is covered by the Plan. This should be a listing of the specific crisis events that are covered, as well as what parts of the organization may need to participate in the implementation.

- **Situations and assumptions.** This section addresses the planning environment and the circumstances under which crises may arise that are covered by the Plan. It also describes how a crisis is identified by internal members of the organization, when the Plan is activated, and what resources are available for each of the crisis events covered.
- **Audience profile.** This section details the different audiences for which the Plan was developed, delineated by specific crisis if needed, and should provide a basic description of each of the audience's characteristics.
- **Concept of operations.** This section describes the general plan of action during a response and makes up the critical functionality of the Plan. Organizational chains of command are identified here as well as the method by which the Plan is activated. Crucial approval processes for what, how, and when communications may occur need to be detailed here. Some Plans use this space to identify the process by which partnership activities are initiated and/or how the organization might begin to become involved in a public response and participation in an emergency operations command center.
- **Key communication strategies.** This section details the methods of communication that will occur both internally and externally and how those channels might be accessed.
- **Organization and assignment of responsibilities.** This section addresses the delegation of responsibilities. This includes the Plan's participants, the designation of a lead spokesperson and additional spokespeople if needed, how 24/7 staffing will be ensured, how to ensure that news is properly disseminated, and how interagency activities will begin to occur. Finally, this section often details, by organization and title, other participants in the response. For ease of understanding, a flow chart or table of organization can be utilized in large, more complex organizations to clearly delineate lines of authority ad relationships between all of the key parties.
- **Evaluation of the communication efforts.** An organization should identify in advance how it intends to evaluate the success of the communication activities when the Plan is implemented so that a method for data gathering can also begin upon implementation of the Plan. This section should detail critical data points. Note that this section is differentiated from the following one in that it collects data for predetermined evaluation measures that will be used both during the Plan implementation as well as after. The information gathered will also be used for Plan development and maintenance, as described in the next category.
- **Plan development and maintenance.** This section provides details on when and how the Plan will be reviewed and updated, including the circumstances under which the reviews will occur. It should also address the methods for controlling the distribution of the document and the method for tracking the reviews through some type of log or similar document.
- **Appendixes.** While the text of the Plan is the working document and should be brief and user-friendly, additional information to be used as references during

implementation need to be readily available and literally attached to the Plan. Some of the more common appendixes include:

- Logistical details
- Call down lists
- Available materials
- Location of prepared key messages (message maps)
- News conference guidelines*
- Media relations reminders*
- Equipment, supplies, and services

*The reader is referred to Chapter 7 for a more detailed discussion of the development of a media communications plan.

REFERENCES

Covello, V.T. and M.W. Merkoher. 1993. *Risk Assessment Methods: Approaches for Assessing Health and Environmental Risks.* New York: Plenum Press.

Fearn-Banks, K. 2007. *Crisis Communications: A Casebook Approach*, 3rd ed. Mahwah, New Jersey: Lawrence Erlbaum Associates.

Haight, J.M., ed. 2008. *The Safety Professionals Handbook: Technical Applications.* Des Plaines, IL: The American Society of Safety Engineers.

Hansen, M. 2008. "Applied Science and Engineering: Systems and Process Safety." In *The Safety Professionals Handbook: Technical Applications*, edited by J.M. Haight. Des Plaines: IL: The American Society of Safety Engineers, pp. 37–69.

Lundgren, R.E. and A.H. McMakin, A.H. 2004. *Risk Communication: A Handbook for Communicating Environmental, Safety, and Health Risks*, 3rd ed. Columbus, OH: Battelle Press .

Manuele, F. 2008. *Advanced Safety Management.* Hoboken, NJ: John Wiley & Sons, Inc.

Manuele, F. 2009. "Prevention Through Design: Guidelines for Addressing Occupational Risks in Design and Redesign Processes." Technical publication TR-Z790.001. American Society of Safety Engineers, Des Plaines, IL.

Manuele, F. 2010. "Acceptable Risk." *Professional Safety* 55(5):30–38.

Reynolds, B. 2002. "Crisis and Emergency Risk Communication." U.S. Centers for Disease Control and Prevention, Atlanta, GA.

U.S. Department of Defense. 2002. "Standard Practice for System Safety." MIL-STD-882D and E. Washington D.C. Also available online at https://acc.dau.mil/CommunityBrowser.aspx?id=255833

9

SPECIAL RISK AND CRISIS COMMUNICATION SITUATIONS

The numbers of risk and crisis situations to which the principles outlined in this text can be applied are infinite and unique to each organization. However, two are selected in this chapter for further study and analysis, in part because the likelihood of their occurrence in most organizations is higher, and also because they present interesting ways in which to illustrate how the concepts in the previous chapters can be applied. The chapter will begin with a review of several concepts introduced previously and then further discuss Sandman's points on dealing with worst-case scenarios, before moving into a discussion of fatalities and rumors.

CRISIS COMMUNICATION PRINCIPLES

As noted in Chapter 2, risk communications and crisis communications differ, primarily with regard to the types of situations in which they occur. Risk communication is an interactive process between an organization and an audience that takes time and resources in order to be effective. Effective risk communications focus on the importance of the interaction between the communicator and the audience as much as the content of the message. The goals of risk communication events are to solicit input and

Risk and Crisis Communications: Methods and Messages, First Edition. Pamela (Ferrante) Walaski.
© 2011 John Wiley & Sons, Inc. Published 2011 by John Wiley & Sons, Inc.

ideas from the audience in order to work together, typically in a problem-solving or consensus-building format. The emphasis in risk communications is on the process and the involvement of the audience, and it occurs when there is no crisis, even though one may be on the horizon.

Risk communication is also an ongoing process that helps to define a problem and solicit involvement and action before an emergency occurs. It is a time-consuming process that involves developing relationships with an audience, sharing information about the nature of risks, and working toward a consensus about the best ways to approach the risk.

Crisis communication has some similarities to risk communication in that it is also an interactive process between a communicator and an audience to transfer information about a crisis. However, the interaction between the parties is less of a focus due to time constraints since the crisis event is already occurring or is about to occur; therefore, the audience members' need for information is more urgent. The goals of the organization in crisis communications are more focused on managing the crisis to lessen the severity of the outcome at the conclusion of the crisis. This can be achieved by convincing the audience members of the need to act in some manner to protect themselves and their interests. Table 9.1 shows the differences between risk and crisis communications.

The application of Sandman's Risk = Hazard + Outrage theory was addressed in detail in Chapter 3, which defined the three variables and illustrated three examples of

TABLE 9.1. Differences between Risk Communications and Crisis Communications

Risk Communications	Crisis Communications
• Event that is the focus of the communications is in the future	• Event that is the focus of the communications is about to occur or is already occurring
• Ongoing process between communicator and audience is time consuming	• Shorter process between organization and audience due to the immediacy of the crisis event
• Focus of efforts is on the dialogue generated between the two parties	
• Most communications are two-way events	• Focus of the efforts is the delivery of messages to the audience
• Goal is to reach consensus with audience regarding activities and solutions to presenting hazard	• Most communications are one-way events
	• Goal is to inform and compel the audience to action intended to keep them safe
• SH&E professional functions include assisting in the risk assessment process to qualify and quantify the risks and assisting in the development of the messages; in some organizations the SH&E professional will also deliver the messages, typically to the workforce	• SH&E professional functions include assisting in the understanding of the severity of the crisis and assisting in the development of the messages; in some organizations the SH&E professional will also deliver the messages, typically to the workforce

crisis communications—precaution advocacy, crisis communications, and outrage management. As noted then and repeated here, in Sandman's view, crisis communications are situations in which the audience outrage level is high. Outrage levels refer to the audience's emotions and include common emotions expressed during a crisis: anger, fear, apathy, mistrust. "Hazard" refers to the actual event that is occurring. Communicating effectively with audiences in a crisis requires a skilled communicator to inform the audience members in a manner that overcomes their emotion so they can hear the message. Additional skills are then required to subsequently craft a message that compels the audience to act in a manner that is believed to help protect it and bring the crisis to a reasonable conclusion.

Communicating in a crisis requires that the audience's legitimate emotions (as noted above, typically fear, mistrust, anger, and apathy) are first acknowledged and understood, even if the level of emotions may be proportionately higher than the situation warrants. (The reader is referred to Chapter 5 for a more detailed discussion of understanding and managing the emotions of fear, mistrust, anger, and apathy.) The communicator must navigate strong emotions, all the while remaining empathetic, rational, and able to demonstrate true leadership. And in some situations, the outrage is directed at the communicator, regardless of the rationality of the blame.

Communicators can improve the process by acknowledging their own emotions about the situation. In fact, in many crises, attempting to hide emotions or pretending they do not exist can serve to increase the fear or add the element of anger to the emotional state of the audience. In addition, messages that attempt to over-reassure the audience, who can clearly see the hazard is dangerous, are not only ineffective, but often increase the level of emotion, making the messages less likely to be heard and understood.

Crisis communications must follow a process of first acknowledging the outrage, then informing and requesting action, and it must be done in that order. Attempts to do otherwise are likely to be unsuccessful. Sandman provides further guidance in his writing about worst-case scenarios and in the process of "moving the seesaw."

WORST-CASE SCENARIOS

Every crisis situation has a potential "worst-case scenario." Sometimes, the actual crisis that is occurring and being addressed in the communication event is the "worst case." More often than not, as a crisis begins to unfold or in the midst of it, the audience is keenly aware of how bad it can get. In fact, the thought of such a scenario can dominate audience's thought patterns, regardless of how imminent the scenario may be, thus causing its outrage level to increase. In turn, communication events become unproductive, or worse, more dangerous to the audience because members are not able to receive the crucial messages about what is happening and how they can protect themselves.

Organizations cannot afford to ignore the worst-case scenario. Crisis communications events are more likely to fail if it is not addressed. Sandman addresses this concept when he says:

A worst-case scenario that your company or agency risk managers see as too unlikely to deserve much attention is likely to strike the public—especially the outraged public—as deserving a lot of attention. If you fail to address the issue or address it too casually or hyper-technically or (worst of all) mockingly, that will increase people's outrage, which in turn will increase their sense that the worst case isn't all that unlikely— launching a cycle of increasing concern you can't afford to ignore (Sandman 2004, p. 14).

In risk terminology, worst-case scenarios are typically high severity/low probability events or event outcomes. During an actual crisis they represent the worst that could possibly happen if the crisis is unable to be managed or if various negative factors all converge or if efforts to manage the crisis fail. During risk communications, these situations often produce ambivalence in the audience, or sometimes just passing concern. There may be small factions of the audience who tend to want to talk about the worst case, but until presented with the actual situations or the precursors to it, the audience typically doesn't allow its outrage level to get out of control thinking and worrying about it.

In addition, as a general rule, audiences tend to be more concerned about catastrophes that produce exceptionally dangerous situations or results than they are about the chronic risks that may ultimately cause their demise. An example of these phenomena is the fear that is generated when a shark attack occurs with resulting substantial injuries (amputations of limbs mostly) or death. Knowledge of a recent shark attack, regardless of the distance, creates an unease in anyone who may be living or vacationing near an ocean, and many beach goers express a nagging worry that may even keep them from swimming in all but the shallowest of waters. In fact, according to the International Shark Attack File (ISAF), there have only been slightly more than 1,900 shark attacks around the world, so that the actual odds of being attacked by a shark are 1 in 11.5 million (ISAF n.d.). However, the risk of suffering a stroke and surviving with complications or dying from one is far greater: 795,000 Americans suffer a stroke every year, and stroke is the number three cause of death (American Stroke Association 2010). Yet, unless an individual audience member has had a recent direct involvement with a stroke victim, the fear of strokes ranks far below that of a shark attack, and despite public education campaigns, many Americans do far too little to reduce their risk of having a stroke.

At first glance this dichotomy may seem irrational, and in fact, many risk managers dismiss it or are scornful as they attempt to convince the audience of the "real" level of risk. But, as Sandman suggests:

That's not because we're stupid, not because we don't understand the data, not because we can't multiply and not because we are misperceiving the risk. It is because we share a societal value that catastrophe is more serious than chronic risk. Possible catastrophes gnaw at people's lives; actual catastrophes are intrinsically unfair and hard to recover from (Sandman 2004, p. 14).

What risk and crisis communicators can take from the above is that when a crisis begins or is imminent, the audience's outrage level begins to rise and the emotions of fear, anger, and mistrust begin to develop and rise accordingly. If not addressed or addressed improperly by the communicator, the level is likely to increase and often will spread in gatherings of audience members as they read about and watch more coverage about the situation in the media.

One additional point should be taken from the example above. An audience's seeming unwillingness to take steps to protect itself from mostly voluntary risks appears to be more related to our assessment of the probability more so than the magnitude. In other words, the risk of dying from a stroke is not something that many people think about on a regular basis; more often it becomes part of their awareness only when it has touched their lives recently. And the further an audience member is from a direct result of the risk, the less likely it seems to be to happening to them. But when the audience member is closer, say vacationing at the ocean, the fear is more strongly perceived. Further, the assessment of risks that are imposed on the audience as opposed to being accepted willingly by our own behavior is also different. As noted in Chapter 4, the perception of the ability to control whether or not the risk directly affects the individual audience member and/or whether they can remove themselves from its path affects the audience and tends to focus its thoughts on the magnitude of what might happen, rather than the probability. In the "moving the seesaw" process discussed below, this concept is crucial to understanding how to craft messages (Sandman 2004).

DEALING WITH AN OUTRAGED AUDIENCE IN A CRISIS

Combining the concepts above provides a communicator with a number of basic principles that are useful for crafting and delivering crisis messages. Underplaying the potential serious effects of a crisis is a technique that risk managers often see as strategically sound in order to protect the audience to keep it from behaving in ways that may appear uncontrollable or to keep it from panicking (see Chapter 5). Along the same lines, risk managers and organizational representatives may believe it is wiser to deal with the worst part of a crisis only if it actually occurs. The rationale behind this strategy is often threefold (Sandman 2004):

1. The assumption is that the audience doesn't need to know. In reality, it is a very rare situation in which the public doesn't need (or deserve) accurate information about events that might impact their lives. In almost every situation, informed audiences can be expected to behave in ways that will benefit and protect them, and it is rare that there isn't something organizations want the audience to do to get ready in the event of a worst case, no matter how unlikely.

2. They reason that the likelihood is so low that describing it in detail will increase panic in the audience or increase their level of fear and anger rather than reduce it. (Chapter 5 provided a detailed discussion about the rare occasions when an audience panics.) The concern is that an audience who takes a risk more seriously than is desired by an organization can cause worry, leading to a belief within the organization that the outcome of honest communication about the

risks will cause more work on the part of the organization to manage the audience's response to the crisis.

3. Some organizational representatives simply believe it is wrong to speculate and that the audience deserves only true and accurate facts about what *is* happening, not details and information about what *might* happen. However, Sandman argues that nearly all risk and crisis communications are speculation. He suggests that the essence of the communication process is telling an audience what the organization *believes* is true and what it *believes* will happen given its assessment of numerous factors. Interspersed within the messages are factual details, but the bulk of the message is already made up of speculative information and associated details.

Both underplaying the potential risks and not mentioning them at all fails for two key reasons: the first is that the audience is almost always already aware of what the worst outcome might be or at least some semblance of it. The audience is probably waiting to see how the organization plans to handle the discussion or it may simply be waiting for the organization to fail to reveal it and then use the lack of openness to demonstrate the organization's lack of compassion and concern. The inability or unwillingness to acknowledge the "elephant in the room" only makes it grow in importance among an audience that is already aware of its existence. This leads to increased frustration on the part of the audience, increasing its lack of trust in all of the other credible messages that are being communicated to it. It is far better for the representative of the organization to acknowledge early on what is known about the worst case and what the organization has spent time and resources planning for, revealing to the audience those plans clearly and without any downplaying. If the event never materializes, the audience will be more forgiving if the outcome never actually occurs or is less serious than is suggested.

The second reason for the frequent failure of this strategy is because, while not likely in terms of pure probability, worst-case scenarios do occur, or perhaps come closer to occurring. Audiences caught unaware by a truly significant crisis that affects their lives in a major way are given to high levels of anger and mistrust if they were never given information about what could happen. The level of outrage produced in this situation further impairs their ability to take the action desired by the organization. Even more problematic is that the audience is often left with substantial residual feelings of anger and mistrust that can affect the organization's relationship for many years to come.

DEALING WITH AN AMBIVALENT AUDIENCE IN A CRISIS

As discussed above, in a true crisis many members of the audience are given to expressing high levels of outrage stemming from their focus on the high severity side of the risk assessment. Whether or not the level is commensurate with the true nature of the hazard depends on many factors, but ignoring the outrage by either downplaying the crisis with the audience or pretending that the really frightful outcomes don't exist creates potentially hostile and explosive situations. These circumstances are best dealt with by the organization by first acknowledging the seriousness of the high severity, low probability events.

However, not all audiences will be highly outraged in a crisis. Sometimes, ambivalence is the predominant emotion, and either audience members see both sides of the hazard or are so outraged by the hazard that they are initially unable to take a position. In the latter situation, in order to settle their anxiety, audiences often choose to focus their attention on the side of the hazard that everyone else is ignoring. For example, if an organization is attempting to reduce the worry level of their audience, they tend to tell them not to worry. Audience members, in seeing that the organization is not worried, decide there must be a reason for them to worry and take that position. Alternatively, if an organization is trying to increase the outrage level of an audience's members and tells them to worry, the audience chooses to do just the opposite.

This tendency creates somewhat of a seesaw, according to Sandman, as the audience attempts to balance the equation by moving to the side that is not being weighted down; the balance tends to resolve various levels of worry and concern. Sandman calls this "moving the seesaw," and moving the audience to the desired outrage level requires a unique and somewhat counterintuitive strategy. Many organizations try to play "follow the leader" with their messages, fully expecting the audience to do exactly, but often the audience does the opposite (Sandman 2001).

To provide clarity to this strategy, Sandman uses the concept of a playground seesaw. A seesaw stays balanced when both sides have equal weight; otherwise one side of the seesaw lies on the ground and the other remains in the air. Without additional weight being placed on the side in the air, nothing can happen. In situations where there is both high severity and low probability, the balance an organization wishes to achieve comes from being able to help the audience see both sides of the issue and take a position that is similar to the way the organization views it.

Sandman's recommendation for achieving this goal says that an organization should take the public position in its messages that is the opposite of what it wants the audience to focus on. If the organization is more concerned about the high severity, it should talk about the low probability; in other words push the seesaw down to the ground on the low probability side. The audience, seeing the high severity part of the seesaw up in the air, moves in that direction to balance its perception of the situation. Alternatively, if an organization wishes the audience to focus on the low probability, it should focus on the high severity, thus prompting the audience to once again take the "empty seat" on the seesaw.

Once the audience moves to the side of the seesaw the organization wishes, more messaging is needed to keep the seesaw balanced. This part of the strategy involves focusing on the long-term outcomes of the risk communications by helping the audience to resolve its outstanding issues with both sides of the equation. By focusing its messages on both the high severity and low probability of the hazard and then theoretically "moving" toward the fulcrum of the seesaw, the audience, wishing to remain balanced, moves back toward the middle as well. Figure 9.1 illustrates Sandman's "move the seesaw" concept.

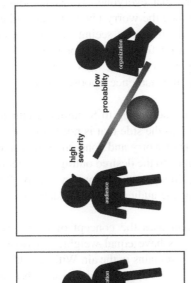

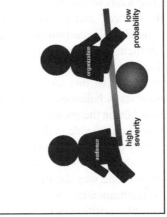

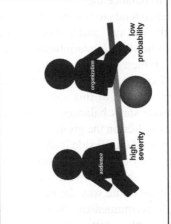

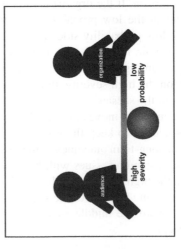

Fig. 9.1. Dealing with Ambivalent Audience in a Crisis—Moving the Seesaw

SOME ADDITIONAL GUIDELINES

In addition to the above methods for dealing with worse-case scenarios, the guidelines listed below provide an additional framework with which to evaluate crisis situations and craft messages (Sandman 2004).

1. **Don't ignore the "best-case" scenario either.** Focusing on the worst-case scenario is important for all of the reasons listed above, but the content of the message can be tempered with "best-case" or "not-so-worst-case" scenarios. These can be an effective means of helping the audience to understand that most crisis situations are a matter of degree and that various events, some within the organization's control and some not, determine where on the continuum the situation eventually lands. Messages that outline the most likely scenario, as well as the ones the organization is best prepared for, along with a few points about how bad things can get, provide a sound set of messages.

2. **Make it clear the future being discussed is possible.** Worst cases are predictions and are simply outlining how bad the risk or crisis situation could be. If the audience is going to understand that concept, it needs to be reinforced throughout the messages.

3. **Keep repeating both sides of the situation—high magnitude and low probability.** Make sure that statements about the worst case also contain messages about the probability. Don't separate the messages into different statements or different communication events.

4. **Make sure the audience knows that the organization is prepared, even if it isn't likely to happen.** Delivering worst-case messages should also be accompanied by what the organization has done/is doing to prepare for it.

5. **Give the audience something to do.** As was noted in Chapter 5, "action binds anxiety." Make sure that the message includes some aspects of what the audience can do to prepare as well. Moving the outrage level from high to low includes both acknowledging the audience's outrage level but also providing action steps it can take to prepare, which will continue to reduce its outrage level and keep it there.

DEALING WITH FATALITIES

In the United States alone, in an average year, 5,000 workers lose their lives in workplace incidents. While many organizations never have had to face dealing with a crisis of this type, many others do, and some have experienced it more than once. Most organizations have emergency response plans with varying levels of response designed to provide guidance in events ranging from fires or explosions to weather-related emergencies to hazardous substances releases. Often, however, these plans neglect dealing with a fatality, leaving many SH&E professionals floundering in the midst of what is an emotional situation for them as well, leading to decision making and actions that are rife with errors. Considering a fatality as a "worst-case scenario" helps to provide the necessary guidance for crafting messages and communicating with other members of the workforce, organizational management, family members, the media, and other interested parties. Considering Sandman's Risk = Hazard + Outrage concept, it would be expected that the outrage levels of the various audiences will be very high, though probably none more so than coworkers of the victim.

Crafting messages for this highly charged emotional situation utilizes some of the same concepts outlined above, including understanding and acknowledging the outrage level, not assuming that the audience doesn't need or want to know unpleasant details, and not underplaying the information presented (Clausen 2009; Sandman 2004). Below is an application of the general concepts to this particular situation:

- Any attempt to downplay the situation will be met with justifiable anger and mistrust. Attempts to over-reassure the workforce that the situation should not be cause for sadness and anger, particularly in the early stages, are at the best ineffective and at worst, callous. The messages delivered early in the event are not easy to deliver and are not pleasant to hear, but the organization must find a way to trust the workforce to act appropriately if given accurate information. Members of the workforce need to be helped to bear their pain and sadness; trying to reduce it too quickly will probably be met with resistance and increased anger and frustration (Claussen 2009). As the crisis ebbs and the workplace returns to a pre-incident state, messages that try to moderate strong feelings are more appropriate, but not in the early hours, days, and even weeks following the incident.

- Returning to concepts detailed in Chapter 3, the communicator will be relying heavily on previously built levels of trust and credibility as the initial messages are delivered. It goes without saying that a workforce with a fundamentally sound level of trust in its leadership will be more receptive to messages during this type of crisis than one in which the relationship between management and the workforce is adversarial. Furthermore, the messages delivered by the organization during the management of this type of crisis can either strengthen the trust and credibility or deteriorate it both during the crisis on beyond.

- Once the message has been delivered, detailing the basic information about what has happened, members of the workforce will often benefit from being given some task that they can perform. As noted above, action binds anxiety, and it can be assumed that the workforce will be suffering high levels of anxiety in the early minutes, hours, and sometimes days following a fatality. It may be that they witnessed the event or it may be that the victim was a close coworker. It may also be that information about the incident is not completely understood, leaving the rest of the workforce to wonder about their own personal safety moving forward. The types of tasks that can be given to the workforce vary and are dependent upon the situation at hand, but some effective ones include assistance with the investigation, making suggestions on corrective actions to prevent a recurrence, and providing help and support to the victim's family and to coworkers. It is also helpful to remember that giving audience members options and choices about the actions they take helps them assume control over both the situation and their own response to it.

- Along the same lines, much was noted in Chapter 5 about an organization's reluctance to provide details to an audience due to concern about panic. It should be restated here that panic in an audience is not a typical reaction, even in the face of extreme danger. When given accurate information and actions to be taken to protect themselves and others, most audience members will do what is in their best interests and, for the most part, what is in the best interests of the group. Failing to follow the recommendations of a communicator is not panic; it is more likely that the audience has heard the information, come to a different conclusion, and taken action based upon its own assessment of the situation and the best course of action.

- Communicators should be instructed to be willing to share their own pain during communication events that discuss the fatality. Acknowledging the personal impact the situation has had on the individual communicator does not make them less professional or ethical; it makes them more human. Few audiences will see that as a negative situation. Furthermore, this type of sharing helps create a "teamwork" mentality on the part of the audience that can impact future relationships not only during the aftermath of the crisis but in the future as well. The bonds that are created when human beings work together to come through a crisis can be strong and deep.

- More will be noted in the next section about rumors in the workplace, but in the event of a fatality, it is critical that communications occur early and often, even in the absence of much information about what happened. The vacuum created by lack of timely information can be detrimental as workers will fill that vacuum with their own ideas and information. Having to counteract false understandings and opinions can be difficult. Even more so is that the absence of information creates the impression that the organization has something to hide or does not think the workforce deserves to know anything and finds its position on the situation unimportant, once again unfavorably impacting the issue of trust and credibility.

- As noted in Chapter 7, the choice of a good spokesperson is essential in any risk and crisis communications event. However, when a fatality occurs, the ability to express compassion and concern is heightened. While many communication situations require someone who can translate complicated concepts into understandable messages, the demeanor of the spokesperson in a crisis like a fatality is more important, particularly in the early stages. What the workforce needs is a spokesperson who can verbalize its outrage, be it anger or sadness or both, and acknowledge that they are understandable emotions. Spokespeople with soft, slow, and compassionate voices along with ones who are able to establish and maintain good eye contact are the best choices in the first few communication events. It may be necessary or desirable to change to a different spokesperson as the investigation moves forward, particularly if understanding the event involves complex and technical details, but the first few events will benefit greatly from a spokesperson whose "human-ness" shines through when they speak.

• While this chapter delves more deeply into crisis communications situations, the long-term communication process and messages that follow a fatality are also important to consider. More is being learned about the circumstances under which post-traumatic stress disorder (PTSD) may occur; the possibility of seeing this type of reaction exists along the continuum from mild to severe among members of a workforce. As soon as the communication events regarding the incident begin, concurrent actions should also begin to provide for the psychological mental health of the workforce and associated audience members. It may be that some members of the audience are unable to leave the worksite and get home on their own and will need escorted. It may be that in the future days, weeks, and months, symptoms of PSTD will emerge. The organization needs to take steps to assure the audience that assistance is available and make the utilization of it simple and private.

DEALING WITH RUMORS

Rumors often occur independently of a crisis and then eventually precipitate it. Furthermore, most crises also involve some spreading of rumors that swirl about the various audiences as the organization attempts to address them. Both scenarios provide rationale for crisis communicators to be prepared to address them, as will be discussed in this section. Regardless of the genesis of rumors, treating their existence as a crisis and elevating their seriousness within an organization to prompt some type of action are crucial. Many an organization has allowed the circulation of rumors to go unchecked and paid the ultimate price.

The effect of rumors on the operations of an organization is many faceted and involves both internal and external aspects. On one end of the spectrum are those minor rumors that cause the audience to be focused on issues other than the organization's mission and goals. If the audience is the workforce, it often becomes distracted, angry, frustrated, and fearful, creating a work environment that is not conducive to producing a quality product or service. Even more problematic is the manner in which a rumor can create a workforce that is more at risk for incidents that produce injuries, property damage, or environmental damage, as is evidenced by the following quote from a worker employed by a aluminum smelter plant where massive workforce cuts as a result of the recession of 2007 were occurring across the company's many plants. Aluminum smelter plants, dangerous workplaces under the best of conditions, require a workforce that is paying attention to the work environment and is rigorous in following safe work practices.

> You can imagine if every day you wake up you go to work, but you don't know if that's the last day you're going to go to work. Some part of you is going to be watching what you're doing. But you're not going to be. . . . Your mind's not totally committed to focusing on what you're doing, when in the back of your head you're thinking, "Man, am I getting laid off? I wish they'd tell us something. All we hear is rumors."

The reactions of other audiences to rumors can also be problematic for an organization. The production and eventual selling of any type of consumer product requires an audience (the consumer) who believes in the basic safety and quality of the product. Rumors that undermine the consumer's confidence in the product can wreak havoc on sales.

The worst-case outcomes from the broad circulation of rumors is when an organization ceases to exist, typically because the publicity surrounding it leads to such a severe reduction in sales that the organization is no longer viable or because of lawsuits, which result in substantial debt from defending the organization or from losses stemming from verdicts against the organization. Although not as drastic, some companies simply never regain their market share and though they continue to remain in business, their size and profits are permanently reduced.

Kathleen Fearn-Banks (2007) addresses the topic of rumors by first identifying rumors by type. She lists the following six types, while also noting that many rumors will overlap categories:

1. **Intentional.** This rumor is intentionally started by an organization with a purpose and can be the type of rumor deliberately started to achieve sales growth, such as is often done by those in the financial industry to increase the prices of stocks or mutual funds.

2. **Premature fact.** In the beginning, these rumors do not appear to have a basis in fact, but eventually do turn out to be true, regardless of how they may seem when they are first started. Workforce layoffs are a good example of this type of rumor.

3. **Malicious.** Like intentional rumors, these are deliberately started, but the purpose is to harm, and unlike intentional rumors, these are started by a competing organization. Political campaigns use this type of rumor fairly often.

4. **Outrageous.** The content of this type of rumor is essentially unbelievable. However, those who end up hearing it may end up assuming that it must be true, since it couldn't be made up.

5. **Nearly true.** Portions of this type of rumor are true, but not all of it, unlike premature fact rumors, where the entire content appears untrue. However, this type of rumor continues to circulate and gain credibility so that those hearing it will eventually decide that the whole rumor must be true since at least part of it clearly is.

6. **Birthday.** These rumors continue to emerge over and over again, just like a birthday. The regularly occurring rumor about Chinese restaurants cooking with dogs or cats is an example.

Regardless of the type, rumors pose unique problems for any organization that go beyond traditional crisis communication messages, and because it is nearly impossible to determine how they start, it is often complicated to discover how far they have spread; careful consideration is required to develop a plan of action. Organizations that do not plan ahead for rumor control are the ones that find themselves spending substantial amounts of time and money on them, sometimes successfully, sometimes not. In this time of rapid communications, most members of an organization's various audiences get their news from television; an endless source of reporting is available on hundreds of channels. In addition, the majority of Americans are also well connected through the Internet, where the delivery of "information" is not always factual nor without ulterior motive; it is unlikely that laws will be enacted to prevent blatantly untrue or misleading information from being published on Internet sites in the near future. Finally, studies currently show that the most "trusted" source of information by people is other people, making rumor control and counteraction an uphill battle, albeit one an organization needs to fight at times to remain in business.

It goes without saying that many rumors never gather traction and disappear before they spread and create problems for an organization. However, those rumors that do become widespread do so because of several key reasons (Fearn-Banks 2007):

- There is something factual or believable in the rumor's content. As noted above, even the most outlandish rumor can begin with a portion of it that is true and accurate.
- Hearing a rumor generally causes strong feelings in the receiver of the message. If the rumor has some validity or applicability to the receiver's life circumstances, those feelings are generally anger, fear, or something equally as strong. Passing the rumor along to another person can ease the stressful feelings and can also help the message sender calibrate its authenticity. If the receiver is a trusted person, their response may help either confirm or deny the rumor's content in the sender's eyes. Depending upon the response, the level of feelings may either dissipate or increase.
- Passing along a rumor can also serve to elevate the sender's authority or level of importance, either in the perception of the receiver or in the sender's self-perception.
- Many rumors portray an organization in a negative light. If the receiver of the message already has negative feelings about the organization, hearing the rumor helps to reinforce their perceptions, regardless of the level of truth in the rumor or the seeming outlandishness of its content.

Dealing with rumors requires some proactive planning, and organizational risk management activities should focus to some degree on the organization's ability to prevent rumors as well as detect them early on. Strategies for both scenarios should be outlined in risk management and crisis management plans. Fearn-Banks (2007) provides some general strategies, along with the caution that dealing with rumors is not an exact science, and each scenario needs to be evaluated with regard to its own particulars.

- Organizations can be the last to know about a rumor or find out about it after it has been circulating for quite some time, thereby limiting the strategies available to deal with it. Rumor hotlines or other mechanisms for publics to provide information, typically confidentially, to an organization can help with early detection.

- Running periodic focus groups that query a variety of audiences about their perspective of an organization can turn up information generally known and accepted by the audience but not the organization. The knowledge might portray the organization in a negative light and eventually lead to a rumor or already be a minor rumor, in low circulation.

- Management of other organizational representatives should be trained in rumor detection and control. Front-line supervisors are often the recipients of valuable information and hear rumors either directly from the workforce or in bits and pieces as they perform their job duties. Training for this portion of the workforce should focus on rumor identification as well as on crafting and delivering by supervisors at the time rumors are spreading to help counteract them and reduce their reach. In addition, supervisors should be given clear guidelines as to when to notify someone in the organization about information circulating that may become a rumor, or already is, so that proactive activities can be moved higher up in the organizational structure.

- Organizations should make it a practice to pay close attention when a similar type of organization is the subject of a rumor and observe how the organization manages it, as well as the outcome of the strategies employed. These types of prodromes provide valuable information that can be used in future efforts. A file of the incident including copies of news coverage and notes about the rumor recipient organization's actions should be maintained within the organization for study. This valuable information provides a ready method of evaluating and possibly revising an existing crisis communications plan framework or for immediately accessible strategies to be employed when a similar rumor affects it directly.

- The lines of communication with key publics should be well developed, allowing for information to flow out when possible problems erupt or are on the horizon. An organization's workforce, as is noted in the quote above, is particularly susceptible to rumors when sufficient information is not provided to help the workforce understand what is happening, to consider how the organization's actions may affect them, and to evaluate what options exist for dealing with the situation.

- An organization should ensure that it has a good deal of trust and credibility with its key audiences. As has been noted in previous chapters, these two factors more than anything else provide an organization with the foundation upon which risk and crisis communications rest. In other words, "The organization must be more credible than the rumor" (Coombs 1999).

All of the information presented above focuses on how an organization can proactively reduce the potential that a rumor will develop. Once a rumor is detected, a different set of strategies is warranted (U.S. DHHS 2006):

- Time is a critical factor in rumor response; a previously developed plan with a set of organizational goals and objectives, spokespersons, and critical message templates will jumpstart any response. The longer the rumor is permitted to persist, the harder it will be to counteract and since rumors have a tendency to morph rapidly, staying ahead of them becomes nearly impossible, as most of the effort will be spent being reactive rather than proactive.
- The organization will need to be able to quickly ascertain the extent of the spread of a rumor, with the understanding that literally each hour that goes by increases the chances that the rumor will spread. This is important to know because the extent of the spread will help determine the proper response. Not every rumor demands a full frontal attack. In some cases, attracting attention to a rumor only increases the chances that it will spread. If it can be determined that the rumor is confined to a small group, working directly with that group quietly and quickly may be the best solution. However, for rumors that are found to be widespread or have the capacity to become so quickly, a more aggressive response in probably in order.
- Along the same lines, effective rumor control occurs when the organization accurately predicts how a rumor might evolve. While the ability to be absolutely accurate is nearly impossible, knowing your audiences through focus groups, open lines of two-way communications, and a risk communications strategy that is regularly implemented and evaluated will provide information, thereby increasing the chance that the organization will not only be able to determine where the rumor is headed but also the types of messages the audience is most likely to respond positively to when the messages are delivered.
- An organization needs to ensure that messages delivered to counteract or dispute rumors are clear and focus only on factual information. Messages that are limited in their opportunity for interpretation and that limit discussion and debate are more likely to be successful. Crisis communications in this type of situation are intended to be one-sided presentations of factual information, not opportunities for dialogue and discussion, as might be true in risk communications.

Other rumor communication strategies are offered by Fearn-Banks (2007):

- Messages delivered regarding rumors should offer information that is contradictory to the rumor's content without ever mentioning the rumor. It is believed that doing so only gives the rumor more credibility and also creates a situation where an audience member may not yet have heard the rumor, thus inadvertently increasing its spread. Audience members who are now aware of a rumor may stop to consider it first, thereby ignoring the delivered messages, which are designed to counteract it. They may then take the additional action of spreading the rumor still further.
- There are some situations in which doing nothing after becoming aware of the existence of a rumor is an effective strategy. This would be true if denying the rumor will only bring more attention to it. As noted above, when a rumor is circulating within a small audience, publicly discussing the rumor only brings it

to the attention of a larger group. Dealing with the rumor quietly only among those who have, or are likely to have, heard it is more effective. One note of caution with this strategy: It is vital to monitor the rumor carefully to see that it begins to die off. If there are signs that indicate the rumor continues to circulate or is beginning to grow, being more aggressive is in order.

- The opposite strategy of loudly and very publicly denying the rumor works best when the organization has proof that is easy for the average person to verify that the rumor is not true. It is also helpful when taking this route to ensure that a positive relationship with the local media is already in place, as they will not only provide a method for spreading the organization's response but can also be expected to verify the offered facts in their reports.

- In order to distance the organization from the rumor, a credible spokesperson from outside the organization, a celebrity or well-known public figure such as a university professor, can be called upon to discredit the rumor in some public manner (e.g., news conferences, media interviews). This strategy is helpful when the organization wants to provide information contrary to the rumor without appearing to directly in its public relations efforts, perhaps because the organization's reputation may be problematic or the spokesperson previously used by the organization was ineffective.

- Advertisements or notices in publications with widespread circulations or distinct circulations within the audience most likely to be spreading the rumor are effective ways to reach large numbers of people. Caution is urged here, however, as the message must be well crafted. Otherwise, this strategy may only serve to make more people aware of the rumor's existence, causing it to spread further. This strategy works best with rumors that have little or no basis in fact and are easily able to be counteracted with facts.

Table 9.2 details possible organizational responses when it has discovered a rumor.

TABLE 9.2. Possible Organizational Responses upon Discovery of a Rumor[a]

Have a plan in place	Time is critical when an organization becomes aware of a rumor. Having a plan in place reduces the amount of time necessary to decide what to do, even if the organization decides not to take any action. Planning allows for proactive action, not reactive.
Be able to ascertain its spread	Having an established network of detection systems allows an organization to determine the extent of the spread of a rumor, an important factor in determining actions to be taken or not taken. Counteracting a rumor within a small group at the earliest stages increases the success of the effort.
Know your audiences	An organization with a sound understanding of the positions of its various audiences can not only be better able to predict where and how the rumor might spread but also can better target crisis messages to achieve the desired result.

(Continued)

TABLE 9.2. *(Continued)*

Limit messages to factual information	This type of communication is not risk based, but crisis based, and therefore it is more reliant on one-sided presentations of the facts in a manner that persuades the audience to accept the organization's position and act accordingly.
Never mention the rumor	Messages designed to counteract a rumor should not mention it specifically but be limited to factual information that counters it. Mentioning the rumor or its contents only serves to further its spread.
Maybe it's best to do nothing	A rumor that is confined to a small number of audience members may die on its own. Speaking of it to a larger group of audience members spreads it. If necessary, work to counteract the rumor only within the audience members who know of it and monitor it. If it spreads, widen the strategies to counteract it.
Make a lot of noise	For rumors that are widespread and fairly easy to prove incorrect, splashy messages designed to draw attention are best. But make sure the relationship with the audience and the media is already positive or this strategy may backfire.
Consider a go-between	If an organization wants to speak against a rumor indirectly, credible spokespersons (celebrities) may work. Their pre-existing reputation improves the chances that the message will be believed by the audience, and the organization can put some distance between itself and the rumor.
Advertise the message	Putting the message out to the audience in a print or verbal advertisement allows an organization to speak about it directly by presenting easily confirmable facts while still maintaining some distance between itself and the message designed to counteract the rumor. Distance achieves the goal of dispelling the rumor indirectly.

[a]U.S. DHHS 2006; Fearn-Banks 2007

REFERENCES

American Stroke Association (ASA). n.d. Posted online at http://www.strokeassociation.org/presenter.jhtml?identifier=1200037. Accessed on August 6, 2010.

Claussen, Lauretta. 2009. "After the Incident: How to Deliver the Message to Employees and Family Members About Workplace Victims." *Safety+Health* 180(5):48–51.

Coombs, W.T. 1999. *Ongoing Crisis Communications: Planning, Managing, and Responding.* Thousand Oaks, CA: Sage Publications, Inc.

Fearn-Banks, K. 2007. *Crisis Communications: A Casebook Approach*, 3rd ed. Mahwah, New Jersey: Lawrence Erlbaum Associates.

International Shark Attack File (ISAF). n.d. Posted online at http://www.flmnh.ufl.edu/fish/sharks/isaf/isaf.htm. Accessed August 6, 2010.

Sandman, P. 2001. "Advice for President Bartlet: Riding the Seesaw." Posted online at www.petersandman.com/col/westwing.htm on July 14, 2001. Accessed on October 23, 2010.

Sandman, P. 2004. "Worst Case Scenarios." Posted online at http://www.petersandman.com/col/birdflu.htm on August 28, 2004 . Accessed on August 6, 2010.

U.S. Department of Health and Human Services. 2006. "Communicating in a Crisis: Risk Communication Guidelines for Public Officials." Washington, D.C.

Ydstie, John. 2010. "Extreme Downsizing May Hurt Companies Later." National Public Radio. Transcript of broadcast posted online at http://www.npr.org/templates/story/story.php?storyId=129036823. Accessed August 9, 2010.

10

CASE STUDIES

This chapter will continue the process of the application of principles begun in the previous chapter by analyzing two recent crisis communications situations: the H1N1 pandemic of 2009–2010 and the BP Deepwater Horizon oil spill of 2010. These particular case studies were chosen for a variety of reasons, not the least of which is the substantial amount of media coverage of numerous communication events that offer the opportunity to analyze the messages and activities of the various organizational representatives. In addition, they provide an increasing level of applicability: the BP Deepwater Horizon oil spill had far-reaching implications throughout the United States and for oil and gas drilling worldwide, and the H1N1 pandemic was a global event. Finally, they provide a variety of differing responses on the part of the principle organization impacted by the crisis. BP management repeatedly spoke of their willingness to pay for all legitimate damages caused by the spill while maintaining the company's unwillingness to accept responsibility for the event. The World Health Organization (WHO) and various United States agencies including the Department of Health and Human Services (DHHS) attempted to both educate and warn an apathetic audience by appearing to create a larger and more serious health crisis than what appeared to be the case by many audience members.

Risk and Crisis Communications: Methods and Messages, First Edition. Pamela (Ferrante) Walaski.
© 2011 John Wiley & Sons, Inc. Published 2011 by John Wiley & Sons, Inc.

Case studies are valuable tools for the study of risk and crisis communications as they allow organizations not impacted by a particular event to clearly view the tremendous problems associated with delivering messages in the current media culture that provides constant access to a variety of images and reports. Crisis management in the current environment places an organization in a seemingly impossible position as it attempts to both accept responsibly and limit liability, as key management listens in one ear to in-house or external media relations specialists and in the other to legal counsel. As members of the organization work to deal with the immediate needs of the crisis, while simultaneously trying to find answers to what happened, it all occurs under the relentless and hypercritical eye of the current media culture, which requires constant access to key organizational representatives for extended periods of time and is quick to pounce on even the smallest error by an exhausted spokesperson who may be dealing with the crisis 16 to 18 hours a day for weeks or months on end.

While it would seem that an organization faces an uphill battle just to stay afloat, the lessons learned by applying basic principles of risk and crisis communications provide much for other organizations to use in the future if they end up in the unfortunate position recently experienced by BP, WHO, and DHHS.

THE H1N1 PANDEMIC OF 2009–2010

A pandemic is "an outbreak of a disease that initiates simultaneous infections of humans throughout the world" (American Industrial Hygiene Association 2006). Pandemics involve disease-causing viruses to which the majority of the human population is vulnerable because they have not previously been exposed to the specific circulating strain and, therefore, have not had the opportunity to build up immunity to it. Pandemics also involve viruses that spread easily from human to human, infecting large numbers of those exposed to it, killing previously healthy individuals at rates not typically experienced from seasonal viruses, overwhelming medical services, and causing massive disruption to social and governmental institutions. However, an important distinction not often understood by the average public audience members is that pandemics are defined by their ability to spread disease, not to cause death.

Some recent historical pandemics include the Spanish Flu of 1918–19, which killed over 500,000 people in the United States and 20 million worldwide; the Asian Flu of 1957–58, which killed 70,000 Americans, mostly elderly and infants; and the Hong Kong Flu of 1968–69, in which approximately 36,000 Americans died, again mostly elderly and infants. Of recent concern to most health officials and the World Health Organization (WHO) is the potential of the H5N1 virus (avian flu) to become a pandemic. Currently circulating mostly in Asia, the virus is transmitted from bird to bird and occasionally from bird to human, with an extremely high mortality rate of up to 55 percent. The concern among public health officials is that if this virus mutates and becomes easily transmissible from human to human, a full-blown pandemic is highly likely to occur with case fatality ratios that will make the effects of the pandemic exceedingly dangerous and comprehensive.

The U.S. government recognized the need for a comprehensive national pandemic plan as far back as 1977 when the Carter administration directed the Department of Health, Education and Welfare (DHEW) to create a National Strategy for Pandemic Influenza (hereafter referred to as the "Strategy"). The draft Strategy, developed by DHEW in 1997, began circulating in an attempt to develop a consensus, particularly among public health and elected officials. DHEW became the Department of Health and Human Services (DHHS) in 1980 and continued to try and develop a consensus on the draft Strategy until the Bush administration brought the planning process back into the White House and assigned the Department of Homeland Security (DHS) the task of finalizing the Strategy and developing an implementation plan. The first finalized plan to be released by DHHS in 2005 was called "The National Strategy for Pandemic Influenza," and in 2006, the Implementation Plan (hereafter referred to as the "Plan") was released (U.S. House of Representatives Committee on Homeland Security 2009).

Shortly after Barak Obama took office, the U.S. House of Representatives Committee on Homeland Security made oversight of pandemic preparedness one of its priorities and requested that a study be undertaken to determine the current status of the Strategy. The report, titled "Getting Beyond Getting Ready for Pandemic Influenza," was released in January 2009 and was sharply critical of the Bush administration, suggesting that there were no areas where the Strategy was complete and rating most areas as inadequate. The report said, in part:

> Work is clearly underway to prepare for such a biological event. However, despite the fact that we are overdue for an influenza pandemic and that we fear the consequences of such a disease spreading unchecked—we are not prepared as a nation to fully withstand the impact of such a devastating widespread biological event. Pandemic influenza would destroy the security of our nation and homeland (U.S. House of Representatives Committee on Homeland Security 2009).

Within months, on April 13, 2009, the predictions made in this sweeping statement would be put to the test when the first resident of Oaxaca, Mexico, died of what would initially be called "swine flu." Figure 10.1 shows the H1N1 virus.

Involvement of Stakeholders in the Strategy Planning Process

Several criticisms of the Committee on Homeland Security's report pointed to the inability to reach a consensus among key audiences of the initial draft Strategy and also of the lack of key audience involvement in the planning process of the Strategy and Plan. As was discussed in Chapter 2, the process of risk communication is one that requires an ongoing partnership and dialogue in order to develop a consensus (as much as is possible). This definition, first noted in that chapter, summarizes the interactive nature of the process:

> Risk communication: an interactive process of exchange of information and opinion among individuals, groups, and institutions; often involves multiple messages about

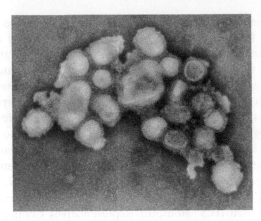

Fig. 10.1. H1N1 Virus (formerly known as the Swine Flu Virus)

the nature of risk or expressing concerns, opinions, or reactions to risk messages or to legal and institutional arrangements for risk management (U.S. DHHS 2006).

The interactive process described above was first begun in 1977 by DHEW and eventually taken over by DHHS. The draft Strategy remained stagnant during the lengthy attempt to gain consensus among the other major federal departments (the audiences in this risk communication effort) by DHHS for more than 25 years. While communicating with audiences about risks is an interactive process that takes time and resources in order to be effective, 25-plus years is clearly an indication of a failed effort, most likely due to the vagaries of the political process and the inefficiency of large bureaucracies. This lack of commitment and inability to finalize a task by government was discussed in Chapter 3 as a key measure of its inability to generate trust and credibility among its audiences and a frequent reason for the lack of success of risk and crisis messages. Following the publication of their landmark study on environmental risk communication, Peters *et al.* (1997) demonstrated that governmental institutions were best able to increase their levels of trust and credibility of their audiences by demonstrating commitment to a process or to a goal. Overcoming this perception among the audience and demonstrating a sincere ability to commit to an effort or project over the long haul are likely to generate the largest change in audience perceptions of trust and credibility. It is likely that the inability of DHEW/DHHS to achieve consensus on the Strategy among their colleagues in other federal departments in the substantial span of time taken is indicative of its inability to demonstrate commitment to the participants of the process and to compel them to get involved. From the report of the U.S. Committee on Homeland Security:

> Although the disease itself falls under the purview of HHS, all members of the Executive Branch must deal with the impact any pandemic would have on society. HHS did not have the authority to compel other federal departments and agencies to take actions to prepare for, respond to, or recover from an influenza pandemic—let

alone make them provide speedy review and relevant commentary on the draft strategy. Clearly, the consensus-gaining process utilized by HHS was also very time consuming (U.S. House of Representatives Committee on Homeland Security 2009).

Once members of the Bush administration took over the completion of the Strategy and eventually the development of the Plan, the process clearly moved much faster, and both documents were finalized within three years. And while a national framework for dealing with a crisis as would be presented by a pandemic is one that may not lend itself wholly to an interactive process of consensus building, clearly a plan detailing the implementation of a pandemic strategy does. But, this critical document (the Plan), developed to establish priorities and prepare the country for a pandemic, did not engage key audiences. The Plan details those activities to be undertaken by numerous state and local agencies as well as the private sector and major social organizations. Yet, according to the report, these groups were absent from the process:

> Notably absent from the White House process to develop the Implementation Plan were a number of key stakeholders, including representatives from the states, territories, tribes and localities, as well as the private sector and the international community. Although the Implementation Plan was briefed to some of these stakeholders (or their representatives) shortly after its release, they were not afforded the opportunity to participate in the process to develop the Plan in the first place. As a result, the executive branch did not have their buy-in or support (U.S House of Representatives Committee on Homeland Security 2009).

As expected, a number of the recommendations of the report from the Committee on Homeland Security included involving key stakeholders and improving the efforts of earlier administrations:

> All stakeholders necessary to the accomplishment of a particular goal or set of objectives must be identified and included in planning efforts. Without the stakeholders responsible for carrying out critical activities or controlling mission-critical resources (necessary to the accomplishment of the goals or objectives), the execution of these plans will be met with difficulty (U.S House of Representatives Committee on Homeland Security 2009).

Public Health Education about Pandemics

Even though WHO declared a Stage 6 pandemic on June 11, 2009, the term "pandemic" was not one that would have been found in the vocabularies of the average U.S. citizen five years ago, possibly even as recently as two years ago, just before the pandemic was declared. Despite this lack of public knowledge, concerns about the heavy casualties, toll on medical and other services, and disruption to critical infrastructures had been recognized for quite some time prior to that. Many in the public health community had understood the term and its implications for years, and, as noted above, the federal government began its preparations with written documents in the late 1970s. Even many occupational safety and health professionals, including this author, had been presenting materials at conferences and client meetings, urging organizations to include pandemic planning in their overall emergency response planning process. However, except for very large organizations and those with a greater than average degree of public safety risks (health-care facilities, universities), many were unaware of even the definition of the term, despite the fact that the general assumption that a future pandemic was not "if" it would occur, but "when." The DHHS report "Pandemic Influenza: Preparedness, Response, and Recovery" (2006) states: "Public health experts warn pandemic influenza poses a significant risk to the United States and the world—only its timing, severity, and exact strain remain uncertain."

The DHHS report "The Next Flu Pandemic: What to Expect" (2010) states: "Flu pandemics have happened throughout history. They occur from time to time and some are worse than others. Public health experts say it's not a matter of IF a flu pandemic will happen, but WHEN. We cannot predict when the next flu pandemic will happen."

When the H1N1 virus began to spread throughout Mexico, primarily Mexico City, public awareness campaigns were ramped up to increase knowledge and understanding. But because of the limited understanding of pandemics and their effects among the general public, much of the activity was focused on remedial concepts: the basic definition of a pandemic, the difference between pandemic influenza and seasonal influenza, and what measures the public could take to protect themselves, including practicing personal hygiene and social distancing. Two specific educational efforts in this arena proved to be effective at substantially increasing the public's awareness of pandemics in general as well as how to protect themselves.

The first was the widely promoted and visible "Cover Your Cough" poster (Fig. 10.2) and associated materials that taught members of the public a whole new method of personal hygiene by coughing into their elbows and not their hands, thus minimizing the spread of germs. Prior to the H1N1 pandemic it would have been unusual for the average American to see, let alone practice, this technique, yet now it seems quite commonplace.

The second educational effort was the reduction in the types of personal greetings, such as handshakes and cheek kisses during the height of the outbreak. Most of the members of the general public seemed to become aware of the need to not only protect themselves from potential exposure to the virus but to protect others from being infected by increasing the amount and type of social distancing techniques they engaged in on a regular basis. Media articles advocating such techniques as the "el-bump" to be used as a general sign of greeting began to proliferate. Dr. Sanjay Gupta, CNN's chief medical correspondent wrote on a blog that the "el-bump" was "not as cool as the fist bump, but safer" (Wong 2009). Even churchgoers were urged to skip the traditional "sign of peace" handshakes, hugs, and kisses and instead consider bowing (Sostek 2009).

Also widely promoted as a source for information about pandemics in general and the H1N1 pandemic specifically was the DHHS website www.pandemicflu.gov, which features many excellent resources aimed at increasing the public's knowledge and understanding of pandemics in general, including the differences between influenza and pandemic influenza, as shown in Table 10.1.

Stop the spread of germs that make you and others sick!

Cover your Cough

Cover your mouth and nose with a tissue when you cough or sneeze

or

cough or sneeze into your upper sleeve, not your hands.

Put your used tissue in the waste basket.

You may be asked to put on a surgical mask to protect others.

Clean your Hands
after coughing or sneezing.

Wash with soap and water
or
clean with alcohol-based hand cleaner.

Minnesota Department of Health
625 N Robert Street, PO Box 64975
St. Paul, MN 55164-0975
651-201-5414 TDD/TTY 651-201-5797
www.health.state.mn.us

Minnesota
Antibiotic
Resistance
Collaborative

ASSOCIATION FOR PROFESSIONALS IN
INFECTION CONTROL AND EPIDEMIOLOGY, INC

IC#141-1428

Fig. 10.2. Cover Your Cough Poster (courtesy of Minnesota Department of Public Health). Used with permission

TABLE 10.1. Seasonal Flu vs. Pandemic Flu[a]

Seasonal Flu	Pandemic Flu
Outbreaks follow predictable seasonal patterns; occurs annually, usually in winter, in temperate climates	Occurs rarely (three times in 20th century)
Usually some immunity built up from previous exposure	No previous exposure; little or no pre-existing immunity
Healthy adults usually not at risk for serious complications; the very young, the elderly, and those with certain underlying health conditions at increased risk for serious complications	Healthy people may be at increased risk for serious complications
Health systems can usually meet public and patient needs	Health systems may be overwhelmed
Vaccine developed based on known flu strains and available for annual flu season	Vaccine probably would not be available in the early stages of a pandemic
Adequate supplies of antivirals are usually available	Effective antivirals may be in limited supply
Average U.S. deaths approximately 36,000/yr	Number of deaths could be quite high (e.g., U.S. 1918 death toll approximately 675,000)
Symptoms: fever, cough, runny nose, muscle pain; deaths often caused by complications, such as pneumonia	Symptoms may be more severe and complications more frequent
Generally causes modest impact on society (e.g., some school closings, encouragement of people who are sick to stay home)	May cause major impact on society (e.g., widespread restrictions on travel, closing of schools and businesses, cancellation of large public gatherings)
Manageable impact on domestic and world economy	Potential for severe impact on domestic and world economy

[a]U.S. DHHS 2010

These and other successful risk communication efforts, while not done prior to the pandemic declaration in June 2009, did succeed in increasing the public's awareness of pandemics and their potential impact. National polls conducted by the Harvard Opinion Research Program at the Harvard School of Public Health and Research and reported in the *New England Journal of Medicine* found substantial increases in the behavior of the public to several targeted messages about personal hygiene, particularly the need to wash hands more frequently, use hand sanitizers, and to "cover their cough" by using their elbow or a tissue. As stated by Steelfisher *et al.* (2010): "Polls during the 2009 H1N1 pandemic also suggest that public health communication efforts related to other personal influenza-prevention behaviors were effective in reaching a large swath of the public."

The Harvard polls found the following:

- In a poll conducted on April 29, 2009, 59 percent of those surveyed reported responding to the flu outbreak by washing their hands or using hand sanitizer more frequently (Blendon 2009a).
- In a poll conducted on June 22–28, 2009, that number was 62 percent (Blendon 2009c).
- In a poll conducted November 12–19, 2009, that number had risen to 82 percent (Steelfisher et al. 2010).
- In a poll conducted on May 5–6, 2009, 14 percent of those surveyed reported that they personally had stopped shaking hands with people and 12 percent reported that they had stopped hugging and kissing close friends and relatives (Blendon 2009b).
- In a poll of behaviors of the traveling public, the percentage of respondents who said they would be certain to sneeze into the elbow rather than on their hands on a future trip was 81 percent, compared to 64 percent who indicated they took the same precaution on a previous trip (Harvard School of Public Health 2009).

How Bad Is a Pandemic Really?—Reducing Trust and Credibility

While many public health risk communication successes can be added to the list above, one difficult aspect of communicating about pandemics relates to the actual definition of a pandemic. As noted above, the term pandemic refers to the *spread* of a disease, not the *severity* of it. Risk and crisis communication messages, such as those presented below, do not take the time to provide clarity to the general public about this critical aspect of the definition. Dr. Margaret Chan, director general of WHO, stated: "After all, it really is all of humanity that is under threat during a pandemic" (Harris and Altman 2009).

From the U.S House of Representatives Committee on Homeland Security (2006):

With the change in administration, there is an opportunity to renew federal efforts to protect our country against all enemies, foreign and domestic. These enemies include infectious diseases, such as pandemic influenza—an enemy perhaps more fearsome than any human adversary. No military force would rest knowing that it was not yet ready to fight the enemy that it knew was advancing to attack.

Dr. Chan was the sole person at WHO responsible for declaring a worldwide pandemic using the WHO pandemic scale, used for the first time during the H1N1 pandemic. The scale (Fig. 10.3) refers only to transmission of a pandemic, not its severity. As was widely reported by the media, Dr. Chan raised the level to 3 on April 25, 2009, to 4 on April 27, 2009, and to 5 on April 28, 2009. In a matter of four days, the public was called upon to absorb the implications of these declarations. When, on June 11, 2009, Dr. Chan declared the H1N1 virus to be a pandemic, it was nearly impossible for the general public, who were, for the most part, learning a whole new language, to

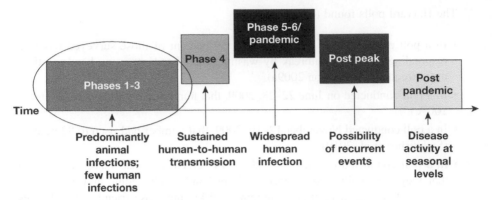

Fig. 10.3. World Health Organization Pandemic Phases

TABLE 10.2. Centers for Disease Control
Pandemic Severity Index[a]

Category	Case Fatality Ratio
1	Less than 0.1%
2	0.1%–0.5%
3	0.5%–1.0%
4	1.0%–2.0%
5	2.0% or higher

[a]Pandemic Severity Index. Posted online at http://
en.wikipedia.org/wiki/Pandemic_Severity_Index;
accessed on September 26, 2010

evaluate the seriousness of the actual virus, or perhaps even understand how it might impact their daily lives and safety.

> The current survey further suggests that the World Health Organization (WHO) decision to raise the worldwide pandemic alert level to Phase 6 did not dramatically impact Americans' level of concern about their personal risk. Only 22 percent of Americans knew that the WHO had raised the level, and only 8 percent of Americans said it made them more concerned that they or their family would get Influenza A (H1N1) in the next 12 months (Harvard School of Public Health 2009).

Many public health officials would argue that a scale much better suited to an understanding of the severity of a pandemic incorporates the concept of a case fatality ratio (cfr) into its designations, which is the number of deaths out of the total reported cases of the disease. The concept of communicating pandemic severity is the intention of the Centers for Disease Control's (CDC) Pandemic Severity Index, as seen in Table 10.2. This index was originally drafted using concepts similar to those used to classify hurricanes. The index, which appeared in an interim report published by the CDC in 2007, was not utilized by the CDC in its communications during the H1N1 pandemic and is no longer accessible on the CDC's website. As of the development of this text, it is still undergoing revision because the CDC believes that it focuses too heavily on severe pandemics with high cfrs. For a moderate pandemic like H1N1, the CDC believes it would have done little to improve the public's understanding.

The CDC reports that it is searching for a scale that is more descriptive and takes into account many factors such as attack rate, hospitalization rate, ICU admission rates, and virus characteristics (Schnirring and Roos 2011).

Following another WHO declaration on August 10, 2009, that the pandemic was over, WHO was urged to evaluate its current scale and make modifications that would address severity as well:

- Dr. William Schaffner, chair, Department of Preventative Medicine, Vanderbilt University: "The WHO needs a mechanism to dial down the anxiety levels while educating us about the extent of the transmission" (Harris and Altman 2009).
- Dr. Julio Frenk, dean, Harvard School of Public Health: "The WHO will analyze its own response to the swine flu outbreak and adjust its system" (Harris and Altman 2009).
- Dr. Harvey Fineberg, president, Institute of Medicine: "They need that other dimension" (Harris and Altman 2009).
- Leung and Nicoll (2010): "Distilling descriptions of the impact of a complex public health threat like a pandemic into a single term like 'mild,' 'moderate,' or 'severe' can potentially be misleading."

In recent statements, Dr. Chan also seems to indicate that such a modification is being considered. According to Roos (2010):

> She (Dr. Chan) said the WHO's pandemic preparedness guidelines were developed by experts "under threat of the H5N1 avian influenza virus and that gave them the sense that collectively we should plan for the worst and hope for the best." One of the lessons is that perhaps we need more flexibility in our future pandemic planning. We need to be able to have a best-case, an intermediate-case, and a worst-case scenario . . . to allow flexibility and adjustments as we track the evolution of the pandemic.

Over-Reassuring the Public about Vaccine Availability

One of Sandman's key concepts in crisis communications messages is that over-reassuring the public in time of crisis creates problems for both message comprehension and for engaging the audience to take the action desired by communicators. He suggests that it is always easier to come back to the audience and let it know that the situation is not as bad as was feared than it is to tell the audience that it is worse than was originally communicated (Sandman 2010b). This error frequently committed by crisis communicators was evident during the H1N1 pandemic with regards to the amount of vaccine available for the public. One of the public health messages communicated regularly during the pandemic was the importance of everyone getting vaccinated as soon as the vaccine became available. The importance of this action by individual members of the public was part of nearly every public health briefing and major media story beginning as early as May 2009, more than one month before the pandemic was declared by WHO. Early on, the messages began to make predictions regarding the amount of vaccine that would be available, long before the virus strain had been characterized. The messages also began to predict the timing of the rollout of a massive public vaccination program. Many of these messages came from high level governmental authorities in DHHS and were widely reported by major media outlets.

Health and Human Services Assistant Secretary Craig Vanderwagen stated, "But it will take several months before the first pilot lots begin required human testing to ensure the vaccine is safe and effective. If all goes well, broader production could start in the fall. We think 600 million doses is achievable in a six-month time frame from that fall start" (O'Neill 2009).

And on June 11, 2009, the same day WHO declared a Stage 6 pandemic, came this statement from the Dr. Thomas Frieden, director of the CDC: "We can expect to see continued efforts to develop a vaccine and we hope and anticipate that that may be in place by the fall" (CDC 2009a).

As the summer wore on, multiple public health messages designed to increase the public's awareness of flu care and treatment—about how to avoid contracting the flu and how to treat it if infected—began to permeate media outlets. As noted above, these messages were effective at changing the public's behavior toward personal hygiene and social distancing. However, among these messages were continued predictions from numerous sources about the timing of the vaccine's availability and the amount of vaccine expected to be available, all of which created an over-reassured public that assumed they would be able to easily get the vaccine when it was available in October if they wanted it.

As Dr. Anne Suchat (pictured in Fig. 10.4), director of the National Center for Immunization and Respiratory Diseases, CDC, stated:

> Of course it is always risky to say that because influenza vaccine manufacturing is not always as predictable as you would like. And sometimes we get surprises. But at this

Fig. 10.4. Dr. Anne Suchat, director of the National Center for Immunization and Respiratory Diseases, The Centers for Disease Control and Prevention

point we're expecting there to be reasonably large numbers of doses available, and the middle of October is the point that we're looking at in terms of our planning, that we hope to be able to launch a vaccine program. The exact number of doses that we'll have, whether everything will be ready to go, those are things that we really have to be prepared for some surprises around (CDC 2009b).

Dr. Suchat also stated: "There was inadvertent messaging that we were going to have a lot" (Schnirring and Roos 2011).

In a joint teleconference with U.S. Education Secretary Arne Duncan, CDC Director Frieden stated: "Shots will start coming off the production line in time for inoculations to start in mid-October," also stressing that although the vaccine "will not all come off the production line at the same time, we do anticipate ample supply for everyone who wants to get vaccinated" (Smith 2009).

According to Jay Butler, director of CDC's H1N1 Vaccine Task Force:

We're expecting somewhere between 45 million and 52 million doses of vaccine to be available by mid-October. This will be followed by weekly availability of vaccine up to about 195 million doses by the end of the year. Keep in mind those numbers are driven by a number of variables in the manufacturing process (CDC 2009b).

Obviously the three quotes from Suchat, Frieden, and Butler do include a message of caution about potential problems with both the manufacturing process and the ability to distribute as much vaccine as would be hoped. A member of the general public reading or listening to them closely would notice that there were no clear promises made. Members of the public health community, whose understanding the process of vaccine development and manufacturing would enable them to be more capable of tempering their expectations of the actual number that would eventually be available. It seems likely that those messages and others were developed as a means of assuring the public that the government was doing everything it could, and they were communicated as a means of both maintaining a sense of comfort on the part of the public that their needs were being taken care of and reducing the spread of panic and fear about the ability to obtain the vaccine when it was distributed to communities for delivery. It also seems likely that these messages were delivered in an attempt to establish at least some political credibility by the current administration to deal with a public health crisis, since media coverage in this current climate is often about the next election and the ability of the current politicians to remain in office or be unseated by one who uses current events as a means of painting a picture of incompetency in the incumbent.

Unfortunately, the cautions in those messages tended to get drowned out as the public began to key in on two points: the number of vaccines that were being produced and that the vaccination program would be launched in full in October. In DHHS Secretary Kathleen Sibelius's testimony before the U.S. House of Representatives Committee on Energy and Commerce on September 15, 2009, she uses the word "purchased" prior to the number of vaccines (195 million), not "available," but that nuance was probably missed in the media sound bite that would have been published, which is the part of the message the public would have been most likely to hear:

With unprecedented speed, we have completed key steps in the vaccine development process—we have characterized the virus, identified a candidate strain, expedited manufacturing, and performed clinical trials. One-hundred-ninety-five (195) million doses of H1N1 vaccine have been *purchased* (author emphasis) from five manufacturers by the U.S. government (U.S. House of Representatives Committee on Energy and Commerce 2009).

And even more confusing (or frustrating) to anyone paying close attention the words of the various messengers is that just two days following Sibelius's testimony, one of the key communicators regarding the CDC's vaccination program uses the word "available" to play the numbers game, and reduces it to 3.4 million, even though in a press briefing just two months earlier, Jay Butler, director of CDC's H1N1 Vaccine Task Force, had mirrored Sibelius's number of 195 million "available":

> So this [the vaccine] has been able to be developed very rapidly, and we anticipate that it will be available during October. Initially we anticipate that about 3.4 million doses of vaccine will be *available* (author emphasis). Additional vaccine will be available, but 3.4 is the hard number that we have right now (CDC 2009c).

By early October, it appeared that the various messages were succeeding in creating a perception among the public that the vaccine was a beneficial step they should take to protect themselves and their families. In a poll taken just after the above quote by Butler, 40 percent of respondents said they were "absolutely certain" they would get the vaccine for themselves and 51 percent were "absolutely certain" they would get it for their children (Blendon 2009d).

Several weeks later, both DHHS and CDC authorities were forced to begin to backpedal on their vaccine predictions and acknowledge problems in both the availability and distribution.

In an Associated Press story, Kathleen Sibelius, secretary, U.S. Department of Health and Human Services said:

> It's [the vaccine] rolling off the production lines right now . . . ahead of schedule and that's good news. . . . By the end of October we should have a substantial amount available and begin to vaccinate a wider population of folks. . . . The early going is a little bumpy, but we'll have a good supply by October (Associated Press 2009a).

Dr. Thomas Frieden, director of the CDC, from a CDC press briefing:

> We are now in a period where the vaccine availability is increasing steadily, but far too slowly. It's frustrating to all of us. We wish there were more vaccine available. As of Wednesday, there were 14.1 million doses available to the states for ordering, 11.3 million had been shipped, so there are now more than 11 million doses in the community in various places. So there is steady progress getting vaccine out (CDC 2009d).

From Roos and Schnirring (2010): "Back in July, HHS had projected that 120 million doses of vaccine would be ready on October. But in mid-August that was trimmed to a predicted 45 million doses by mid-October."

In the first message, Sibelius leaves out numbers altogether as she made the rounds of the various morning news shows in the first weeks of the vaccine distribution, when it became clear that the earlier predictions about numbers were off and that the availability of vaccine was not going to pan out as predicted. In the second message, Frieden expresses his feelings about the inability to get the vaccine distributed as quickly as possible when, after a few weeks of the vaccination program, photographs of long lines and interviews with increasingly frustrated and angry members of the public began to surface. As predicted by Sandman, if the messengers had been less reassuring early on, they would not have had to attempt to explain why things had not worked out as planned. The messages would have acknowledged that the amount of vaccine available was less than what would have been hoped for, but the public would have heard repeatedly that they should not expect much vaccine until the end of the year or early in the beginning of the next year. And unfortunately, presidential adviser David Axelrod and Sibelius resort to blaming the manufacturers for the problem, rather than acknowledging the mistakes of her messages.

David Axelrod stated: "We have 28 million doses as of last week. It's growing every day. We expect to get 10 million this week. And we're catching up quickly. But we did represent to the public what we were told by the manufacturers and that turned out not to be the case" (Inskeep 2009).

From Kathleen Sibelius: "We were relying on the manufacturers to give us their number and as soon as we got numbers we put them out to the public. It does appear that those numbers were overly rosy. I hope that people aren't discouraged. I know it's frustrating to wait in line" (Associated Press 2009b).

While the public agreed with Sibelius that the blame should be placed on the manufacturers, they also did not appear willing to absolve the government of some of the responsibility for the problem. In a poll taken by USA Today/Gallup in November 2009, 62 percent of respondents blamed the manufacturers for the shortage, but 58 percent placed a great deal or moderate amount of blame on the government as well (Sternberg 2009).

Others in the crisis communications community were more willing to voice frustrations with the over-reassuring messages delivered by various communicators and expressed concern that the failure of the messages would not only create a public who would end up not getting the vaccine because of the difficulties with availability, but that it would harm future messages and decrease the credibility of the message communicators, particularly those in the government, whose trust and credibly factors were already low.

In late 2010 a study was undertaken by members of the Center for Biosecurity of UPMC in Maryland. The study authors interviewed national, state, and local leaders to evaluate the H1N1 vaccination program. While study participants were sympathetic to the challenges of vaccine distribution, they were also critical of the lack of coordination.

From Rhambhia *et al.* (2010):

In some instances, poor communication between government agencies and clinicians, unavailable information, and administrative delays hindered the H1N1 response.

Project participants noted that poor communication about the availability of vaccine contributed to a lack of credibility that would later hinder public engagement efforts when adequate vaccine became available.

Another of the critics was Michael Osterholm, director of the Center for Infectious Disease Research and Policy at the University of Minnesota. Below are portions of a transcript from an interview broadcast on November 3, 2009, with Steve Inskeep of National Public Radio (Inskeep 2009):

PROF. MICHAEL OSTERHOLM: So I think many of us were concerned that the projections made earlier this year may have been far too optimistic, given that. But having said that, there is a lot of truth to the fact that no one really knew what the amount would be and no one did anything wrong in terms of not making the vaccine the right way. They've made it safely. They've made as much as they can make, and we're just now living with the 1950s technology that is, unfortunately, terribly unreliable in amount—not in safety, but in amount.

STEVE INSKEEP: Was there any substantive damage done by making it an overoptimistic prediction?

OSTERHOLM: Well, I think there is, in the sense that we still have a significant communication problem with the public. We now have about 30 million doses that have been made. However, even as of Friday, only 16.9 million of that has been shipped. You've heard that we're going to need two doses in children under the age of 10, and there's at least 150 million Americans that probably want this vaccine, and a large number of those want it right now.

If you think about the fact that we're ramping up at about 10 million doses a week for the next few weeks, that means that there are many Americans who will not have access to this vaccine well after Thanksgiving, while the disease continues to ramp up now. *I think you're going to see a collision course kind of scenario here in the next two to three weeks as more and more Americans want the vaccine, have this perception because of the messaging that there's a lot of it out there and not be able to get it, and I think you're going to see some real angry people out there. And we've got to start to better communicate to them what the issues are, how much is there and when they can really expect to get it* (author emphasis).

INSKEEP: How do you go about that? I mean, the fact is no matter what you tell people, they're going to want the vaccine.

OSTERHOLM: I think that's absolutely the case, but I think what we have to do is just be honest with them and say this is how much we have. This is why we have what we have and why we're trying to get it to the people we are.

The first and most important message is those people who are at the highest risk for having the severest disease: pregnant women, individuals with asthma, people who are overweight, particularly those obese. People like that really need to get this vaccine first, because it's not that the rest of us won't get disease, but they are much more likely to have severe disease.

I think the second thing that we have to do is get the message out then how to get this vaccine effectively. What we're seeing happen now is that people in their almost panic-like mode are calling medical clinics every day, they're calling pharmacies every day, and we're starting to see phone lines tied up where we can't do routine business. People who have other conditions are having a hard time getting through to the medical clinic.

We've got to get those people out of that scenario. We've got to get them into phone lines where they can call and know where the vaccine's at and communicate more clearly so that I know when my chance is. Part of the problem right now is people just don't know when it's going to be there. They just keep turning to the media: there's lots of vaccine out there, which is, as we just discussed, is not the case yet.

©2003 National Public Radio, Inc. Excerpt from NPR® news report titled *Projection For Swine Flu Vaccine Was Too Optimistic* by NPR's Steve Inskeep was originally broadcast on NPR's *Morning Edition*® November 3, 2009, and is used with the permission of NPR. Any unauthorized duplication is strictly prohibited.

And from Peter Sandman (2009):

It doesn't really matter whether U.S. public health officials actually "promised" there would be ample vaccine by the end of October, or implied it, or merely allowed others to surmise it without working to correct the record. Nor does it matter whether officials were trying to reassure the public or were genuinely overconfident themselves. What matters is that most Americans came to expect there would be ample vaccine by the end of October. And there wasn't. When the audience ends up with false expectations, by definition the communicator did something wrong.

As noted, many of the messages about the vaccine's availability and timing included clear warnings about potential problems, but within those messages were numbers and a consistent time frame of availability in October. Those are the key points that most of the general public heard the loudest, and as they began to make decisions about whether or not to get the vaccine when it was available, they made those decisions based upon a message of over-reassurance from the governmental authorities, particularly from the CDC and DHHS. There is no doubt that it is difficult to balance a message of positive news with negative, particularly when the public's concern is being aroused and they are expecting their government to "do something" about it. But when the amount of vaccine failed to live up to initial projections, the public wouldn't have been angry, frustrated, mistrustful, and the government would have not lost some of its hard-won credibility if they hadn't been led, probably inadvertently, to believe something that didn't come to pass. Had officials followed good risk communications principles, they would not have over-reassured the public regarding the availability of vaccines, and their messages would have lent themselves to sound more like those below:

From Jay Butler of the CDC:

We are working hard with our vaccine manufacturers to get as much vaccine produced as quickly as possible. Because there are too many variables that would affect how much vaccine we can expect to be produced and by when, we aren't prepared to give any hard and fast numbers right now. We think we will be able to get somewhere between 45 million and 52 million doses of vaccine by the time the vaccination program is over, but we think that won't be until after the first of the year.

From Kathleen Sibelius of DHHS:

With unprecedented speed, we have completed key steps in the vaccine development process—we have characterized the virus, identified a candidate strain, expedited manufacturing, and performed clinical trials. The U.S. government is prepared to purchase as much vaccine from these manufacturers as we need so that everyone who wants to be vaccinated can be. We will be staying on top of this situation daily to make sure the manufacturers work as quickly as they can and we will be providing them with as much support as we can to help them produce as much vaccine as quickly as they can.

And from Dr. Frieden, of the CDC:

The vaccine will not all come off the production line at the same time we will make sure that every dose gets to the public as quickly as possible so that everyone who want to receive the vaccine will be able to do so, even though it will probably take to the end of the year before that can happen.

The result of the problems discussed above with regard to over-reassuring is not just for the immediate crisis. It is also likely that the public will remember the failings of the messages for some time to come. The public's memory regarding the failings of the vaccination program will negatively affect their perceptions of future messages and their willingness to take the action being suggested by the communicators.

Success of the Government's and Public Health System's Efforts

The issues of trust and credibility raised earlier in this section impact the overall evaluation of the public's perception and response to the government's risk communication efforts. The following quote appeared in a retrospective published one year after the first death from what would come to be called the H1N1 virus; the issue is brought to bear on the environment the government faced at the beginning of the pandemic.

In our own country, the virus struck at a time when Americans seemed particularly skeptical about out government and large institutions. At times, health officials erred in their recommendations. It is not an easy task, but our public health authorities need to become clear about the lexicon of uncertainty (Wetzel 2010).

However, in the early days of the pandemic, and by the end of it, studies appeared to show that the public felt that both government and public health officials had done an even-handed job of warning the public without causing an overreaction. From Steelfisher *et al.* (2010):

Throughout the H1N1 pandemic, more than half the U.S. population appeared to have a positive impression of the government's response. For example, in the early days of the pandemic, 54% believed the response of the federal government was appropriate, whereas 39% believed the government had overreacted. Nine months later, 59%

believed that public health officials did an excellent or good job in their overall response to the pandemic, whereas 39% believed they did a fair or poor job.

From a Harvard School of Public Health press release: "In the view of more than half of the adults (54%), public health officials spent 'the right amount' of attention on the H1N1 flu outbreak . . . and 16% felt they spent 'too little'" (Harvard School of Public Health 2010).

THE BP DEEPWATER HORIZON OIL SPILL

This crisis dominated the front pages of major newspapers and became the lead story on news broadcasts for months beginning with the explosion, fire, and eventual sinking of the Deepwater Horizon Oil rig on the night of April 20, 2010. As this text is being written, the oil leak has been stopped after spewing somewhere in the neighborhood of 140 million gallons of oil into the Gulf of Mexico. (Figure 10.5 shows an open-water oil burn.) The total amount of oil spilled remains disputed for reasons that are further explained below; however, even using the lower of the estimates, the spill quickly overshadowed the size of the Exxon Valdez oil spill in 1989, which leaked a mere 11 million gallons, although it was nowhere near the largest oil spill on record—460 million gallons that were deliberately released by the Iraqi forces during the Persian Gulf War in 1991. (Figure 10.6 shows the many responders to this crisis event.)

Fig. 10.5. May 6, 2010. Open-water oil burn. Photo taken from the EPA's Airborne Spectral-photometric Environmental Collection Technology (ASPECT) aircraft

Fig. 10.6. Many responders come and go to fill specific roles during a response of this size. "Job labeled vests" help identify the role a person plays in a command post where numerous agencies are involved. U.S. EPA photo by Eric Vance

It should also be noted that each major oil spill from the past 25 years is unique in several key characteristics. The Valdez spill was an above-water spill from a grounded tanker. The total potential volume of oil that could spill was never in question and the area impacted by the spill remained fairly small and isolated. But it was the first major oil spill to be so comprehensively covered by the media so it is easily recalled by many members of the general public, usually by simply using the name of the tanker. The basic details about what happened are also fairly simple—a tanker ran aground in a known hazardous area (Bligh Reef) causing a gash in the side of the tanker from which the oil spilled. The BP Deepwater Spill was an underground spill from a nearly completely drilled oil well that had yet to go into production. However, the technical nature of what happened introduced a new set of words and phrases into the vocabulary of most audience members including "blow-out preventer," "drilling mud," "low marine riser package," "top kill," and "junk kill." It also occurred nearly 5,000 feet underwater and was visible only after images from the underwater video camera were made public. The Persian Gulf Oil spill was unleashed deliberately by Iraqi forces (the "enemy"), the damage was contained to an area that the vast majority of Americans would never get close to, and coverage was limited to war footage showing the blazing oil rigs and fires that burned long after the war ended.

It should be noted that, due to the long-standing nature of the BP oil spill and the sheer number of communication events available for analysis, the focus of this case study is limited to the first three months, beginning with the initial communication events that discussed the technical terms and the immediate steps being taken to deal with the spill until the announcement of the reassignment of Tony Hayward, the chief executive of BP plc, and the main spokesperson for BP in the early stages of the crisis.

Profiling the Audience—Risk Perception

How seriously the audience views the crisis or risk is dependent upon how it views a number of key factors. Chapter 5 introduced Covello's model of the 15 key factors vital in understanding how the audience in this crisis perceived the events that were being explained to them through various media outlets. Building on a table illustrating the 15 factors first introduced in Chapter 5, Table 10.3 adds in an analysis of the 15 key factors as they relate to this crisis.

As Table 10.3 demonstrates, the audience's perception of the overall risk of deepwater drilling and the potential for a spill of such magnitude would in general be fairly low. Their fear or dread of such a spill would also be fairly low, unless they were identified with some group that made opposition to deepwater drilling a key issue. Audience members would also be fairly unfamiliar with the technical nature of the risk and the eventual outcome if the hazard occurred. They would be far enough removed from the immediate implications of the risk as to be less disturbed by it.

However, the audience's perception of the above factors created a situation where it was fairly unprepared for a significant spill, and when graphically and daily confronted with it through the media, the response, as expected, was strongly negative regardless of whether or not the spill had any direct impact on the audience's daily lives. In addition, given the length of time it took for the initial oil spill to be stopped and the years it will likely take to return the area to pre-spill conditions, it is fairly easy to foresee an audience that was primed to be angry from the outset of the spill and possessing significant levels of mistrust of the organizational and governmental spokespersons. It is also fairly easy to predict that a few mistaken comments or failed procedures would significantly raise the negative emotional levels of the audience members.

Understanding the Technical Nature of the Spill

In the early hours of the event, it was enough to simply tell the audience that there had been an explosion and fire aboard the oil rig, resulting in the deaths of 11 of the slightly more than 100 workers on board at the time, that the ensuing fire eventually caused the rig to collapse, and that the explosion caused the underwater pipeline at the bottom of the sea floor to leak, sending oil to the surface of the water. As the hours turned into days and then became weeks and months, the level of technicality of the communication events needed to be increased to include information about the process of underwater drilling and the methods that were used to attempt to cap the flow of oil and clean up the resulting mess left behind. As was noted in Chapter 6, the groundwork needed to be laid in the initial communication events through the use of a set of consistent terms and phrases that described what was occurring (U.S. DHHS 2006). This technique appeared to achieve some measure of success as the terms "blowout preventer," "top kill," and "relief wells" were explained repeatedly by management representatives of BP and other companies involved in work on the drilling rig as they described their actions to try and stop the leak.

TABLE 10.3. Applicability of Risk Factors to BP Deepwater Horizon Oil Spill[a]

Risk Factor	Applicability	Applicability to BP Deepwater Horizon Oil Spill
Voluntariness	If the audience members perceive the risk to be voluntary, they are more likely to accept it because they understand their role in experiencing the implications of the risk.	Audience members would have some level of perception about the voluntariness of this risk, as drilling for oil underwater is not necessary except to provide for the amount of oil most audience members voluntarily consume. For many years, there has been public discussion about the need to reduce the amount of oil consumed, although most of those discussions have been framed in the reduction of the reliance of foreign oil, perhaps leaving some audience members to accept this crisis as a necessary part of drilling for self-produced oil.
Controllability	If the audience members perceive that they have control over the risk, they are more likely to accept the implications of it.	Audience members would have had no role to play in determining whether or not this crisis occurred, leading to low acceptance of the event.
Familiarity	If the audience members have some previous knowledge of the risk or experience with it, they are more likely to accept the implications of it because of the increased level of knowing what might or might not happen.	As was noted above, the technical nature of underwater drilling is not familiar to most audience members, many of whom appear to have been unaware of how much oil could possibly spill with a catastrophic failure of an oil rig such as Deepwater Horizon.
Equity	If the audience members perceive the implications and consequences of the risk to be equally shared among audience members, they are more likely to accept the implications of it.	Most audience members would likely understand that the immediate day-to-day life implications are equally shared except for those whose livelihoods are dependent upon the industries in the Gulf of Mexico (fishing, oil production, and tourism). The recent devastation caused by Hurricane Katrina in the same area might lead some audience members to have a sense of unfairness for those who live in the area.

TABLE 10.3. (*Continued*)

Risk Factor	Applicability	Applicability to BP Deepwater Horizon Oil Spill
Benefits	If the audience members perceive the ultimate benefits of the risk to be positive, they are more likely to accept the potential negative implications of experiencing it.	In some respects, as noted above, audience members might understand that the risks of offshore drilling have some negative implications. Their perception of just how risky the drilling might be was probably nowhere near the level of the final outcome of this crisis. The audience might have been more willing to accept the implications of a smaller spill that was readily contained, but it is difficult for the audience to accept the types of negative implications of this spill.
Understanding	If the audience members possess a basic understanding of the risk, they are more likely to accept the implications of it. The greater the level of understanding, the higher the acceptance.	Deepwater oil drilling was debated for many years and was somewhat of a public controversy. The oil companies were able to secure drilling permits by convincing the governmental authorities that the risk of a major spill was very low and that if one occurred, the companies were prepared to deal with it. While subgroups of the audience may have remained skeptical and in major opposition to deepwater drilling, most of the audience went about their daily lives until the current crisis. Now the oil companies are finding it hard to answer the questions about how it could have happened and why they were not better prepared to deal with a major spill.
Uncertainty	If the audience members perceive the risks have a degree of certainty in various dimensions and in the scientific information available about it, they are more likely to accept the implications of it.	It is not likely that most of the members of the audience ever spent a significant amount of time fully analyzing the risk assessment information made available during the debates about deepwater oil drilling so their analysis of the information and the certainty was probably limited as well. They most likely accepted the views of other organizations with whom they identified (e.g., environmental groups in opposition and business groups in support)

(*Continued*)

TABLE 10.3. (*Continued*)

Risk Factor	Applicability	Applicability to BP Deepwater Horizon Oil Spill
Dread	If the audience members' emotions with regard to a risk are less intense and fearful, the more likely they are to accept the implications of it.	Most audience members do not dread an underwater oil spill and would accept the possibility of its occurrence without fear of how it might impact their daily lives. They might be less accepting now, however, as they have seen the impact of a spill of this magnitude.
Trust in institutions	If the audience members perceive the institutions more significantly involved in the risk as trustworthy and credible, the more likely they are to accept the implications of it.	As corporations, major oil companies probably do not have a significantly high level of trustworthiness. The seemingly random fluctuations in the prices of gasoline are often covered by the media and feature a frustrated audience that voices a belief that corporations continue to make a profit at their expense.
Reversibility	If the audience members perceive the risk to have reversible adverse effects, they are more likely to accept the implications of it.	An oil spill of this magnitude has essentially no reversibility and much of the coverage has focused on the unknown (and unknowable) length of time it will take before the gulf returns to pre-spill conditions. Most audience members are likely to have good understandings of the irreversibility of this crisis.
Personal stake	If the audience members perceive the risk to be limited in its personal implications and consequences, the more likely they are to accept the implications of the risk.	As noted above, the vast majority of the audience is not directly affected by the spill, even if they have a great deal of sympathy for the audience members living in the area and the environmental damage that is being caused.
Ethical/moral nature	If the audience members perceive the risk to be morally or ethically acceptable, they are more likely to accept the implications of it.	The perceived lack of ethics among major corporations is not the issue in this factor. Most audience members would not have a strong position either way regarding the ethics or morality of the risk of deepwater oil drilling, except for their consideration of the lives lost the night of the explosion and the risk to fellow audience members who work in the industry.

TABLE 10.3. (*Continued*)

Risk Factor	Applicability	Applicability to BP Deepwater Horizon Oil Spill
Human vs. natural origin	If the audience members perceive that the origin of the risk is naturally occurring, they are more likely to accept the implications of it.	The origin of this risk is completely human and based upon a decision made by humans to take on the risk. However, since many audience members may not have been deeply involved in the controversy surrounding whether or not to drill offshore they were likely to accept the risk before it occurred, though probably not now.
Catastrophic potential	If the audience members perceive that the amount of fatalities, injuries, and illnesses from a risk are minimal, they are more likely to accept the implications of it.	Oil rig explosions are relatively rare events and the consequences of ones that have occurred fairly recently have not been nearly as serious as Deepwater Horizon. The risk now perceived by the audience has been elevated.

[a]Covello et al. 2001 (Table 1: Risk Perception Factors)

Jack B. Moore, president and CEO of Cameron International Corp., testified before the U.S. House of Representatives Subcommittee on Oversight & Investigation on May 12, 2010:

> The Cameron product used by the Deepwater Horizon is called a "blowout preventer" or "BOP," a product that Cameron actually invented in the 1920s that allows our customers to control the pressure in a well while being drilled (U.S. House of Representatives Subcommittee on Oversight & Investigation 2010a).

Steve Newman, CEO of Transocean, Ltd., also testified before the subcommittee on the same date:

> A BOP (blowout preventer) is a large piece of equipment positioned at the top of the wellhead to provide pressure control. BOPs are designed to quickly shut off the flow of oil or natural gas by squeezing, crushing or shearing the pipe in the event of a "kick" or "blowout"—a sudden, unexpected release of pressure from within the well that can occur during drilling (U.S. House of Representatives Subcommittee on Oversight & Investigation Testimony 2010b).

Lamar McKay, chair and president of BP America, testified before the U.S. House of Representatives Committee on Natural Resources on May 27, 2010:

Our first priority is to stop the flow of oil and secure the well. In order to do that, we are using multiple deepwater drilling units, numerous support vessels, and remotely operated vehicles (ROVs), working on several concurrent strategies. . . . Our primary focus over the last week has been on what is known in the industry as a "top kill." It is a technique for capping wells which has been used worldwide. The technique will inject heaving drilling mud into the blowout preventer (BOP) and well bore in an attempt to kill the well. If necessary we are also preparing a "junk shot" technique to clog the BOP and stop the flow. This involves the injection of fibrous material into the BOP (U.S. House of Representatives Committee on Natural Resources 2010).

And within in a brief period of time, the terms became part of everyday language as evidenced by the following headlines:

- " 'Top kill' fails to stop flow of oil, BP says" (CNN.com 2010)
- " 'Top kill' fails to plug leak; BP readies next approach" (Krauss and Kaufman 2010)
- "Americans wait to learn if top kill will stop oil" (Nuckols and Bluestein 2010)

Trust and Credibility

Many pages of the earlier chapters of this text discussed the key concepts of trust and credibility. The level of both that exists at the beginning of the crisis between the communicator and the audience has a significant impact on how the messages are viewed by the audience. A landmark study by Peters, Covello, and McCallum in 1997 questioned whether or not it could be determined *what* increases the audience's level of trust and the credibility of the organization, thereby enabling messages and activities to be more specifically targeted towards those goals. Two key findings apply to this case study; the first is restated here (the second appears later in this section):

Finding #1. In the **industrial sector**, an increase in the audience's *perceptions of concern and care* provides for the largest increase in trust and credibility by the public of the organization. A common stereotype of many industries is the perception that the organization is more concerned about profits rather than people. Therefore an industrial organization that can use risk communication events to develop or increase the audience's level of perception that the organization also cares about what happens in the community is likely to be more successful.

The implications for BP regarding the above finding are that the ability to demonstrate a concern for people over profits and the ability to communicate a certain level of care regarding the implications of the oil spill would increase the levels of trust and credibility on the part of the audience. An analysis of the messages from BP and the company's actions reveal two key aspects of this concept. The section below will begin to look at various messages from key BP executives and the level of concern and care they communicated. The second is in how the difficulties in attempting to stop the oil spill and the length of time it eventually took to succeed worked against BP, a discussion about which appears further below in the section about how much oil was being spilled

Worst-Case Scenario—How Much Oil?

While testifying before the House Committee on Natural Resources on May 27, 2010, Lamar McKay, chairman and president of BP America stated:

> We have all experienced a tragic series of events. I want to be clear from the outset that we will not rest until the well is under control. . . . This was a horrendous accident. We are all devastated by this. It has profoundly touched our employees, their families, our partners, customers, those in the surrounding areas, and those in government with whom we are working. There has been tremendous shock that such an accident could have happened, and great sorrow for the lives lost and the injuries sustained (U.S. House of Representatives Committee on Natural Resources, 2010).

And as part of his testimony before the House Committee on Energy and Commerce on June 17, 2010, BP's Chief Executive Tony Hayward said:

> The explosion and fire aboard the Deepwater Horizon and the resulting oil spill in the Gulf of Mexico never should have happened—and I am deeply sorry that they did. This is a tragedy: People lost their lives, others were injured, and the Gulf Coast environment and communities are suffering. This is unacceptable, I understand that, and let me be very clear: I fully grasp the terrible reality of the situation. I know that this incident has profoundly impacted lives and caused turmoil, and I deeply regret that (U.S. House of Representatives Committee on Energy and Commerce 2010).

Hayward also stated in his testimony: "Let me be clear: BP has accepted this responsibility and will fulfill this obligation. We have spent nearly $1.5 billion so far, and we will not stop until the job is done" (U.S. House of Representatives Committee on Energy and Commerce 2010).

But was it enough to increase the level of mistrust by the public already in place as discussed above in the section on audience profiling? In the same testimony, Lamar McKay acknowledged those perceptions by saying:

> But I hear the concerns, fears, frustrations—and anger—being voiced all across the country. I understand it, and I know that these sentiments will continue until the leak is stopped and until we prove through our actions that we will do the right thing. Our actions will mean more than words, and we know that, in the end, we will be judged by the quality of our response. Until this happens, no words can be satisfying (U.S. House of Representatives Committee on Natural Resources 2010).

However, further analysis of the messages delivered by key BP executives frequently led to an uncomfortable combination of both accepting responsibility and, at the same time and in the same message, suggesting that the incident was not the fault of BP and that they were merely *victims of an unpreventable accident*, as well as only one of many companies that shared some of the blame. From the same testimonies of Hayward and McKay came the following messages:

Tony Hayward:

> The investigation team's work thus far suggests that this accident was brought about by the apparent failure of a number of processes, systems, and equipment. While the team's work is not done, it appears that there were multiple control mechanisms—procedures and equipment—in place that should have prevented this accident or reduced the impact of the spill. . . . The truth, however, is that this is a complex accident, caused by an unprecedented combination of failures. A number of companies are involved (U.S. House of Representatives Committee on Energy and Commerce 2010).

Lamar McKay:

> But the investigation team's work so far suggests that this is a complex accident involving the failure of a number of processes, systems, and equipment. There were multiple control mechanisms—procedures and equipment—in place that should have prevented this accident or reduced the impact of the spill. Put simply, there seems to have been an unprecedented combination of failures (U.S. House of Representatives Committee on Natural Resources 2010).

BP appears to have done a good job at communicating its genuine sorrow for what happened; their executives' words attempt to convey their understanding of just how awful the incident was. But the problem is that everyone feels bad about what happened. Rare would be the person who could consider the ramifications of an incident where 11 people lost their lives, thousands lost their livelihoods, and hundreds of thousands lost their way of life, along with the ability to enjoy their homes and communities for a long period of time, and not have some compassion. What BP failed to do in its messages was take its expressions of sadness about what happened and its willingness to make it right and combine them with a message that also accepted some blame for what happened, instead of attempting to say that the organization was only one of many victims and that some of the other complainers involved in the drilling operations were at fault too. The latter may very well be the case, but BP was seen as the leading company as evidenced by their role as operator of the rig. Other companies that were involved in the incident were subcontractors to BP, and regardless of the roles the other organizations played in the decisions that were made in the lead-up to the explosion, BP had the final say.

In his analysis of BP's crisis communication efforts, Peter Sandman (2010b) said:

> But apologies require more than regret. They require acceptance of responsibility. The Committee (during his testimony) kept pressing Hayward to concede that BP had contributed to the disaster by cutting corners and ignoring red flags. Or, indeed, that BP had done anything wrong, anything that contributed to the disaster, anything at all. He wouldn't go there, insisting that we should all await the conclusions of the many

ongoing investigations rather than speculating on what went wrong. That may be good litigation strategy, but it isn't good risk communications.

In addition to the problems associated with the above messages, an unfortunate misuse of a word led to the backfiring of a well-intentioned message when BP Chairman Carl-Henric Svanberg, late in the crisis, said the following after a meeting with President Obama on June 15, 2010 where he pledged $20 billion in funds to be available to compensate victims of the oil spill: "I hear comments sometimes that large oil companies are greedy companies or don't care, but that is not the case with BP. We care about the small people" (Mohr 2010).

For some members of the audience, Svanberg's clumsy use of the term "small" was an understandable translation error and he quickly attempted to correct it: "What I was trying to say—that BP understands how deeply this affects the lives of the people who live along the Gulf and depend on it for their livelihood—will best be conveyed not by any words but by the work we do to put things right for the families and businesses who've been hurt" (Mohr 2010).

While both remarks were an attempt by BP to demonstrate that it did understand the frustration that the "average" American would feel coming up against a giant multinational corporation and was an obvious attempt by BP to put a human spin on its understanding of the situation, it fell short due to the poor choice of words as well as the unique point in the crisis when the message was delivered. By June 16, 2010, the crisis was nearly two months old and most of the audience was weary with failed attempts to stop the flow of oil.

Finding #2. In the *government sector*, an increase in an audience's *perceptions of commitment* provides for the largest increase in trust and credibility. The common stereotype about governmental organizations is that they lack stability; the political party in power determines the goals and efforts of the organization, which may not always be what is best for the audience. And when election results change those in control of the organization, the commitment to previous causes or efforts may be moved to a much lesser level of priority or even completely ignored. The notion that most politicians are looking out for themselves and their ongoing electability is one that can be seen in voter polls and letters to the editors, among other similar venues. Crafting messages that overcome this perception among the audience and demonstrate a sincere ability to commit to an effort or project over the long haul are likely to generate the largest change in audience perceptions of trust and credibility.

Figure 10.7 shows governmental officials participating in community meetings to address the concerns of Gulf residents.

One of the key tasks then for the Obama administration was to overcome the audience's assumption that the government would not be able to sustain its role in dealing with the oil spill over the long haul, and that in time, its commitment to the clean-up and to holding BP and other organizations responsible, would flounder or end. As this

Fig. 10.7. May 13, 2010. EPA senior officials, along with other responding agency representatives, listen in and speak to the concerns of nearby residents. U.S. EPA photo by Eric Vance

Fig. 10.8. May 2, 2010. President Barack Obama delivers remarks in the rain at the Coast Guard Venice Center in Venice, La. Official White House photo by Pete Souza

text is being written it is too soon to ascertain whether or not this has come to fruition; however, several message examples demonstrate a fairly tightly controlled message on the part of governmental spokespersons. It would appear that the Obama administration did a good job at keeping its message consistent. Figure 10.8 shows President Obama at the Coast Guard Venice Center in Venice, La.

President Obama:

> I'm here to tell you that you are not alone, you will not be abandoned, and you will not be left behind. The media may get tired of this story, but we will not. We will be on your side and we will see this through" (Superville and Loven 2010).

From the president's weekly address of June 5, 2010:

> These folks work hard. They meet their responsibilities. But now because of a man-made catastrophe—one that's not their fault and that's beyond their control—their lives have been thrown into turmoil. It's brutally unfair. It's wrong. And what I told these men and women—and what I have said since the beginning of this disaster—is that I'm going to stand with the people of the Gulf Coast until they're made whole (Obama 2010a).

The president addressed the nation on June 15, 2010, regarding the BP oil spill:

> Make no mistake: We will fight this spill with everything we've got for as long as it takes. We will make BP pay for the damage their company has caused. And we will do whatever's necessary to help the Gulf Coast and its people recover from this tragedy. . . . I will meet with the chairman of BP and inform him that he is to set aside whatever resources are required to compensate the workers and business owners who have been harmed as a result of his company's recklessness. And this fund will not be controlled by BP (Obama 2010b).

Ken Salazar, secretary of the Interior, is quoted in *USA Today*: "They will be held accountable. We will keep our boot on their neck until the job gets done" (Jervis 2010).

In an opinion piece in *USA Today*, Salazar stated: "BP and everyone who is responsible for this catastrophe will be held accountable for their actions. And they, along with other offshore oil and gas operators will face more oversight, policing and safety standards" (Salazar 2010).

He also stated:

> We are fighting the battle on many fronts. At the president's direction, his entire team will not rest until the oil spill is stopped, the clean-up is completed, and the people, the communities, and the affected environmental are made whole. Let me be very clear: BP is responsible, along with others, for ensuring that the flow of oil from the source is stopped; the spread of oil in the Gulf is contained; the ecological values and near shore areas of the Gulf are protected; any oil coming onshore is cleaned up; all damages to the environment are assessed and remedied; and people, businesses, and governments are compensated for losses. From day one my job has been to make BP and other responsible parties fully accountable. . . . And while the investigations as to the root causes are still underway, we will ensure that those found responsible will be held accountable for their actions. . . . The President has been clear: we will not rest until this leak is contained and we will aggressively pursue compensation for all costs and damages from BP and other responsible parties (Salazar 2010).

From Janet Napolitano, secretary of DHS: "We are going to stay on top of this and stay on top of BP until this gets done and gets done the right way" (Jervis 2010).

The earliest news reports that followed the sinking of the oil rig began to speculate about how much oil had spilled and might spill by the time the crisis could be resolved. One of the first hurdles to overcome was the lack of knowledge by the majority of the audience about the typical unit of measurement for oil—the barrel, which is 42 gallons. Initial reports of the amount of oil being spilled each day had to be regularly calibrated in both barrels and gallons, and it continued to be reported that way for quite some time. It is difficult to determine whether or not this confusion was ever resolved among the majority of the audience.

What proved to be more difficult was finding a way to get all of the major groups involved in the spill to agree upon how much was leaking each day and what the running total was. Competing interests regarding how much oil was leaking confounded any attempts to achieve such a consensus. Environmental groups were motivated to promote estimates on the high side, while BP was in the position of promoting a lower estimate (or attempting to deflect a discussion of the amount) because, unfortunately for the company, a portion of the fine that would eventually be paid to the federal government was statutorily based upon the amount of oil spilled. Making a determination that could be used to calculate the fine would be a difficult task at best because it involved calculating the amount of oil leaking from a broken pipe 5,000 feet underwater. Using estimates from the underwater video camera and combining that with the amount of oil that was eventually captured at the surface by skimmers and other methods would turn out to be the best method. And although the end result would not be an absolute amount that could be categorically verified, it would result in a fairly sound estimate that could be used to calculate the fine.

The earliest reports in late April and early May pegged the leak at a mere 1,000 barrels a day, but after the amount was challenged by an environmental group, it was raised to 5,000. Once the underwater camera video became available for study, the federal Flow Rate Technical Group began to broadcast a range of 12,000–19,000 per day by late May. By early June the same group had again increased the estimate and suggested it was 20,000–40,000 barrels a day. This estimate was increased yet again to 35,000–60,000 barrels a day just one week later in mid-June and remained the last estimate widely used before the leak was finally stopped (Simon, White, and Fausset 2010).

However, for financial reasons noted above, BP was in the unfortunate position of having to try and push the published estimates as low as possible. Any communications from BP executives would certainly be scrutinized when the calculations were being made; therefore, in almost every situation when BP executives were asked about how much oil was leaking, they responded in one of two ways. First they either said their position was that the amount of oil spilled didn't matter because the important issue was getting the leak stopped and the spill cleaned up. While this may appear on its face to be the argument of a company more concerned about people over profits, given the emotional level of the audience, it had the opposite effect, sounding more like a company trying deflect the question, which was exactly what it was doing. The second strategy was to be on the low end of what most scientific experts and governmental representatives were suggesting when BP was willing to provide an estimate. Inevitably, both of these positions caused BP to appear as though it were hiding something and doing so for purposes that ensured the company's profitability rather than its concern for the people affected by the spill.

It was, in many respects, a losing position all around for BP, although Sandman suggests that the company did have a means by which it could both protect itself and satisfy the audience, thereby increasing its level of trust and credibility among the audience, or at the very least, holding the level steady. His recommendation is to always err on the alarming side in crisis communications that are designed to help an audience bear its justified distress. BP's failure to do so early on in the crisis only fueled the controversy about how much oil was leaking, which helped generate accusations that BP was covering it up. Had BP's messages started out with estimates that were in the vicinity of the other experts and not consistently far below them, the audience's level of mistrust might not have been as high. Sandman (2010a) states:

> BP's failure to err on the alarming side, early on, has turned the ongoing uncertainty about how many gallons/barrels of oil are escaping every day into a controversy that has provoked accusations of cover-up. The ordinary citizen hasn't a clue how many gallons or barrels constitute a really bad spill; we don't even know how many gallons in a barrel. We take our cues from accompanying language; "only" 5,000 barrels a day is a lot smaller spill than "as much as" 5,000 barrels a day. And we take our cues from how the numbers are changing. If BP keeps saying 5,000 a day and the other experts start saying 12,000 to 25,000 barrels a day, we're likely to conclude that BP has been intentionally downplaying the spill. That leaves us more distrustful and more alarmed than we would have been if BP had started with an estimate of 25,000 barrels a day. In a crisis it is extremely damaging to come back later and say, "It's worse than we thought." Far better to come back later and say, "It's not as bad as we feared."

Other means of dealing with the estimate of oil would have been to continually focus on the ever-changing nature of the spill as is evidenced by a comment from Coast Guard Admiral Thad Allen, the federal government's incident commander for the spill: "I think we are still dealing with the flow estimate. We're still trying to refine those numbers" (Borenstein and Weber 2010).

As noted above, BP's attempts to stop the oil spill were complex and technical. Their ability to help the audience understand what happened to cause the explosion and fire and what was being attempted to stop the leak are also discussed above; however, BP appears to have missed a key opportunity to demonstrate in a visual message how much it was willing to disclose to the audience when it stonewalled attempts to have the video feed from the underwater camera broadcast live. At the time the issue of whether or not the camera view should be broadcast was being debated at the end of May, the anger and mistrust of the public was beginning to grow. The efforts at stopping the spill via the "top kill" and "junk shot" maneuvers had recently failed, causing an already frustrated audience to become even more so. And even though all along BP had said that the only certain method of stopping the spill was to drill relief wells, a process that would take well into August to be successful, it is likely that most of the audience dismissed those statements in favor of what they were hearing and seeing in media reports about the efforts that were occurring in the immediacy of the crisis.

When the video feed first became available on May 28, 2010 via the PBS "Newshour" program, 1 million viewers tuned in and saw footage that must have been shocking and frustrating as they were now able to clearly visualize how much oil was leaking from the well in a way that numbers of barrels of oil being spilled per day could not provide. Over the next few days, as the video feed was broadcast more widely, numerous websites crashed due to the high volume of viewers attempting to download the video feed. Coupled with BP's initial attempts to prevent the video feed from being broadcast and its ongoing attempts to brush aside the question of how much oil was being spilled, the audience's level of fury only increased. From an Associated Press news story published on May 31, 2010, the following quote illustrates the mood of the audience:

> Faith in institutions—corporation, government, the media—is down. Americans are angry, and they long ago grew accustomed to expecting the resolution of problems in very short order. So when something undefined and uncontrollable happens, they speculate in all the modern forums about collusion and nefarious dealings (Anthony and Foster 2010).

It seems reasonable that BP would do all it could to stop the oil spill as quickly as possible. The delay in doing so caused its eventual fine to increase, the costs of the clean-up to mount as the oil continued to spill, and the value of the company's stock continued its free fall as long as the crisis occurred. Only audience members who truly look for conspiracy theories would suggest that BP was somehow delaying the final stoppage for reasons that would benefit the company. However, BP's position of trust and credibility with the audience was damaged by its inability to appear to be in agreement with scientific and governmental estimates of the amount of oil being spilled as well as its unwillingness, at first, to allow the audience to see the oil spill via the underwater video camera.

Worst-Case Scenario—How Long Until the Leak Is Stopped?

While there are many criticisms that can be drawn regarding the handling of crisis communications about the spill, there is one area where BP, in conjunction with the government, did an effective job. Almost from the beginning of the spill both parties resisted making strong predictions of how long it would take until the leak was finally stopped. In the early days of the crisis what happened and how to fix the leak were essentially unknown due to the uniqueness of the incident and the lack of previous experience in dealing with an underwater spill of the type and magnitude created by the rupture of the underwater pipeline. Most BP executives and government spokespersons pointedly reiterated that concept over and over again. BP and the government articulated fairly early on that while they would put forth numerous efforts to stop the leak all at the same time, all of the efforts, except for the drilling of a relief well (which would take a lengthy period of time), were simply their best attempts to try anything to stop the leak, and therefore they offered no guarantee of success. Resisting some type of guarantee is difficult to do in the midst of uncertainty and relentless questioning about when the crisis will be resolved; a similar concept was illustrated above in the messages that overplayed the number of vaccines and when they would be available during the H1N1 pandemic. Getting caught up in the need to provide guarantees and promises is something Sandman calls "overoptimistic over-reassurance" and always creates more problems in the end, particularly when the over-reassurances fail to materialize (as in the above example of how much oil was leaking).

In early June 2010 when the second attempt to cap the leaking well appeared to be working, both BP's Kent Wells and President Obama voiced their hopeful pessimism, with Wells stating, "I would say things are going as planned. I am encouraged. But remember we only have 12 hours' experience" (Krauss and Fountain 2010).

Obama said, "We are prepared for the worst, even as we hope that BP's efforts bring better news than we've received before" (Krauss and Fountain 2010).

Initial estimates for the completion of drilling relief wells were made in May and June and were projected for late summer, at best. When these estimates were first outlined, much of the audience expressed frustration at the length of time it would take. This frustration created a significant temptation to revise the estimates downward to placate an angry audience. Yet, both BP and the government continued to hold firm. Even in early July, when efforts to stop the leak at the source through the top kill and junk kill maneuvers had not achieved any success, and progress at drilling the relief wells appeared to be ahead of schedule, neither party would budge.

Robert Dudley, BP managing director said, "In a perfect world with no interruptions, it's possible to be ready to stop the well between July 20 and July 27" (Reeves and Breen 2010).

Thad Allen (shown in Fig. 10.9 with President Barack Obama) stated: "There are certain things that can move that date up, but my official position is the middle of

Fig. 10.9. May 2, 2010. U.S. Coast Guard Commandant Admiral Thad Allen, serving as the national incident commander, and EPA Administrator Lisa P. Jackson brief President Barack Obama about the situation along the Gulf Coast following the BP oil spill at the Coast Guard Venice Center in Venice, La. Official White House photo by Pete Souza

August. If it happens sooner than that, I think we can all jump for joy" (Reeves and Breen 2010)

And finally, when the second cap was placed over the leaking well and appeared to be holding, no one was willing to say that the situation was resolved.

Doug Suttles, BP COO: "It's a great sight. It's far from the finish line. It's not time to celebrate" (Long and Weber 2010).

President Obama: It's "a positive sign, but we're still in the testing phase . . ." (Long and Weber 2010).

Choosing a Spokesperson Wisely and Knowing When to Let Them Go

Almost from the beginning, the most frequent voice of BP during communication events was its chief operating officer, Tony Hayward. He endured hours and hours of relentless questioning by the media, had nearly his every waking move filmed and photographed, and undoubtedly spent countless hours and hours working on the crisis. In the enormous volume of quotes from him, several stand out as significant mistakes that caused untold damage to BP and the position it was trying to present to the audience. Given that BP's status as a wealthy multinational corporation with several other major disasters of recent memory (The Texas City, Texas, oil refinery explosion in 2005, and the Prudhoe Bay, Alaska, oil spill of 2006), BP started off at a disadvantage with the audience whose level of mistrust and anger were already in place. And eventually BP did take Hayward out of many of the spokesperson situations, moving him to a different position within the organization shortly after the spill had been capped; it is hard to determine when a spokesperson's gaffe is fatal.

Hayward's first misstatement came in early May when he tried to put into perspective the amount of oil that was flowing from the leak and the amount of potentially toxic dispersant that was being used to break up the oil in relation to the volume of the ocean. While his statement began with a fairly effective attempt to assure the audience that BP would stop the leak, he ended up with a quote that caused great controversy: "We will fix it. I guarantee it. The only question is when. The Gulf of Mexico is a very big ocean. The amount of oil and dispersant we are putting into it is tiny in relation to the total water volume" (Kollowe 2010).

Hayward later infuriated many Gulf Coast residents when he was quoted in an interview suggesting that Americans were likely to file bogus claims for compensation from damages from the spill (Satter and Mohr 2010), and just a few short weeks later, when tensions about the lack of success of the efforts by BP to stop the leak were beginning to rise, Hayward again began a message with an attempt to apologize to Gulf Coast residents, but ended the statement with a comment that created a firestorm of controversy, most likely responsible for his removal from the lead spokesman role and one that will remain firmly entrenched in the minds of many members of the general public: "We're sorry for the massive disruption it's caused to their lives. There's no one who wants this thing over more than I do, I'd like my life back." (*The Times Online* 2010).

The final blow appears to have been not as much about what Hayward said as what he did on June 19, 2010. Very few would dispute that Hayward deserved a day off due to the relentless schedule he had been keeping since mid-April. And just days earlier, BP had made an unprecedented promise to set aside $20 billion to help oil spill victims, money over and above what it was legally required to provide. But when Hayward was photographed attending an exclusive yachting competition in his native England, the level of trust and credibility among the primary Gulf Coast audience spiraled downward and would never recover. Shortly afterward, BP made it official and moved Hayward into a new position within a division of BP based in Russia and appointed Robert Dudley to the position of president and chief executive of a newly formed organization created to manage the spill and its aftermath: BP's Gulf Coast Restoration Organization.

REFERENCES

American Industrial Hygiene Association (AIHA). 2006. "The Role of the Industrial Hygienist in a Pandemic." AIHA Guideline No. 7.

Anthony, T. and M. Foster. 2010. "No End in Sight for Oil Gusher." Associated Press, May 31.

Associated Press. 2009a. "Sibelius: Americans Must Get Swine Flu Vaccination." October 7.

Associated Press. 2009b. "HHS' Sibelius: Ample Flu Vaccine Will be Available." October 26.

Blendon, R. 2009a. "Public Views of H1N1 I." Harvard Opinion Research Program, Harvard School of Public Health. Field date: April 29, 2009. Retrieved at http://www.hsph.harvard.edu/research/horp/project-on-the-public-response-to-h1n1/public-views-of-h1n1-i/index.html.

Blendon, R. 2009b. "Public Views of H1N1 II." Harvard Opinion Research Program, Harvard School of Public Health. Field dates May 5–6, 2009. Retrieved at http://www.hsph.harvard.edu/research/horp/project-on-the-public-response-to-h1n1/public-views-of-h1n1-ii/index.html.

Blendon, R. 2009c. "Public Views of H1N1 III." Harvard Opinion Research Program, Harvard School of Public Health. Field dates: June 22–28, 2009. Retrieved at http://www.hsph.harvard.edu/research/horp/project-on-the-public-response-to-h1n1/public-views-of-h1n1-iii/index.html.

Blendon, R. 2009d. "Public View of the H1N1 Vaccine." Harvard Opinion Research Program, Harvard School of Public Health. Field dates: September 14–20, 2009. Retrieved at http://www.hsph.harvard.edu/research/horp/project-on-the-public-response-to-h1n1/public-views-of-h1n1-vaccine/index.html.

Borenstein, S. and H.R. Weber. 2010. "New Oil Numbers May Mean More Environmental Damage." Associated Press, June 11.

CNN.com. 2010. "Top kill' fails to stop flow of oil, BP says." Posted online at http://www.CNN.com on May 29. Accessed May 29, 2010.

Centers for Disease Control and Prevention (CDC). 2009a. Press Briefing Transcripts, June 11. Accessed online at http://www.cdc.gov/media.

Centers for Disease Control and Prevention (CDC). 2009b. Press Briefing Transcripts, July 24. Accessed online at http://www.cdc.gov/media.

Centers for Disease Control and Prevention (CDC). 2009c. Press Briefing Transcripts, September 18. Accessed online at http://www.cdc.gov/media.

Centers for Disease Control and Prevention (CDC). 2009d. Press Briefing Transcripts, October 23. Accessed online at http://www.cdc.gov/media.

Covello, V., R. Peters, J. Wojtecki, and R. Hyde. 2001. "Risk Communication, the West Nile Virus Epidemic, and Bioterrorism: Responding to the Communication Challenges Posed by the Intentional or Unintentional Release of a Pathogen in an Urban Setting." *Journal of Urban Health* 78(2):382–391.

Harris, G. and L. Altman. 2009. "Managing a Flu Threat With Seasoned Urgency." *New York Times*, May 10.

Harvard School of Public Health. 2009b. "Poll: Travelers Taking Significantly More Precautions Against H1N1 and Seasonal Flu on Trips This Year, Citing Public Health Advice on Sneezing and Hand Sanitizing." Press release, December 10. Retrieved at http://www.hsph.harvard.edu/ research/horp/files/press_release_text_-_h1n1_vaccine_shortage_follow-up.pdf.

Harvard School of Public Health. 2010. "Nearly Half of Americans Believe H1N1 Outbreak is Over, Poll Finds." Press release, February 5. Retrieved at http://www.hsph.harvard.edu/ research/horp/files/press_release_text_-_public_views_of_h1n1_vaccine_-_update.pdf.

Inskeep, Steve. 2009. "Projection for Swine Flu Vaccine Was Too Optimistic." National Public Radio. Transcript of NPR interview conducted by Steve Inskeep with Michael Osterholm posted online at http://www.npr.org/templates/story/story.php?storyId=120044053. Accessed November 3, 2009.

Jervis, R. 2010. "Officials Turn Up Pressure on BP." *USA Today*, June 25.

Kollowe, J. 2010. "BP Chief Executive Tony Hayward in his Own Words." *The Guardian*, May 14.

Krauss, C. and L. Kaufman. 2010. " 'Top kill' fails to plug leak; BP readies next approach." *The New York Times*, May 29.

Krauss, C. and H. Fountain. 2010. "Cap on Well is Reported to Recover 6,000 Barrels of Oil." *The New York Times*, June 5.

Leung, M.L. and A. Nicoll. "Reflections on Pandemic (H1N1) 2009 and the International Response." *PLoS Med* 7(10):e1000346. doi:10.1371/journal.pmed.1000346.

Long, C. and H. Weber. 2010. "Gulf Geyser Stops Gushing, But Will It Hold?" Associated Press, July 16.

Mohr, H. 2010. "Carl-Henric Svanberg Says BP Cares About 'Small People,' Those Hit by Oil Spill React." *Huffington Post*, June 16.

Nuckols, B. and G. Bluestein. 2010. "Americans Wait to Learn if Top Kill Will Stop Oil." Associated Press, May 29.

O'Neill, X. 2009. "Biden: Stay Off Subways During Swine Flu Panic." Associated Press/NBC New York, May 1.

Obama, B. 2010a. Transcript of Weekly Presidential Address, "Speaking From Louisiana on the Oil Spill," June 5.

Obama, B. 2010b. Transcript of Remarks by President Obama to the Nation on the BP Oil Spill, June 15.

Peters, R.G., V.T. Covello, and D.B. McCallum. 1997. "The Determinants of Trust and Credibility in Environmental Risk Communication: An Empirical Study." *Risk Analysis* 17(1):43–54.

Rhambhia, K.J. *et al.* 2010. "Mass Vaccination for the 2009 H1N1 Pandemic: Approaches, Challenges, and Recommendations." *Biosecurity and Bioterrorism: Biodefense Strategy, Practice, and Science* 8(4):321–330.

Reeves, J. and T. Breen. 2010. "Transferring Oil From Broken Well an Option For BP." Associated Press, July 9.

Roos, R. 2010. "WHO Says H1N1 Pandemic is Over." *CIDRAP News*, August 10.

Roos, R. and L. Schnirring. 2010. "More Clouds in H1N1 Vaccine Supply Picture." *CIDRAP News*, October 23.

Salazar, K. 2010. "We're Cleaning Up MMS." *USA Today*, May 24.

Sandman, P. 2009. "U.S Pandemic Vaccine Supply and Distribution: Addressing the Outrage." Posted online at http://www.petersandman.com/col/vaccinesupply.htm on November 18, 2009. Accessed September 27, 2010.

Sandman P. 2010a. "Communicating About the BP Oil Spill: What to Say; Who Should Talk." Posted online at http://www.petersandman,com/articles/deepwater2.htm on June 1, 2010. Accessed June 1, 2010.

Sandman, P. 2010b. "Risk Communication Lessons from the BP Spill." *The Synergist* 21(8):29–31.

Satter, R. and H. Mohr. 2010. "BP CEO's Yacht Outing Infuriates Gulf Residents." Associated Press, June 20.

Schnirring L. and R. Roos. 2011. "CDC Officials Share Key Pandemic Communications Lessons." *CIDRAP News*, January 27.

Simon, R., R. White, and R. Fausset. 2010. "Oil Spill Estimates Revised Upward Again to 60,000 Barrels a Day." *Los Angeles Times*, June 16.

Smith, P 2009. "Swine Flu Vaccine Available in October." *Pittsburgh Post Gazette*, August 11.

Sostek, A. 2009. "Swine Flu Forces Many Adjustments in the Region." *Pittsburgh Post Gazette*, May 2.

Steelfisher, G. *et al.* 2010. "The Public's Response to the 2009 H1N1 Influenza Pandemic." *New England Journal of Medicine* 362:e65(1)–e65(6).

Sternberg, S. 2009. "Poll: Drugmakers Most at Fault for Swine Flu Vaccine Shortage." *USA Today*, November 9.

Superville, D. and J. Loven. "Tar Balls and Promises: Obama Visits Gulf Coast." Associated Press, May 29.

The Times Online (London). 2010. "Embattled BP Chief: I Want My Life Back." Posted online at http://business.timesonline.co.uk/tol/business/industry_sectors/natural_resources/article7. html on May 31, 2010. Accessed on October 23, 2010.

U.S. Department of Health and Human Services. 2006. "Communicating in a Crisis: Risk Communication Guidelines for Public Officials." Washington, D.C.

U.S. Department of Health and Human Services. 2010. "The Next Flu Pandemic: What to Expect." Posted online at http://www.flu.gov/professional/community/nextflupandemic.html. Accessed on September 26, 2010.

U.S. Department of Homeland Security. 2006. "Pandemic Influenza: Preparedness, Response, and Recovery."

U.S. House of Representatives Committee on Energy and Commerce Hearing. 2009. Testimony of Kathleen Sibelius, Secretary, U.S. Department of Health and Human Services, September 15.

U.S. House of Representatives Committee on Energy and Commerce Hearing. 2010. Testimony of Tony Hayward, Chief Executive, BP plc, June 17.

U.S. House of Representatives Committee on Homeland Security. 2009. "Getting Beyond Getting Ready For Pandemic Influenza."

U.S. House of Representatives Committee on Natural Resources Hearing. 2010. Testimony of Lamar McKay, Chairman and President, BP America, May 27.

U.S. House of Representatives Subcommittee on Oversight & Investigations Hearing. 2010a. Testimony of Jack B. Moore, President and CEO of Cameron International Corporation, May 12.

U.S. House of Representatives Subcommittee on Oversight & Investigations Hearing. 2010b. Testimony of Steven Newman, Chief Executive Officer, Transocean, Ltd., May 12.

Wetzel, R. 2010. "What We Learned From H1N1's First Year." *New York Times*, April 13.

Wong, G. "Flu Virus Sparks 'Social Distancing' Trend." Posted online at http://www.cnn.com on May 1, 2009. Accessed on May 1, 2009.

11

SUMMARY AND CONCLUSIONS

In order to provide concluding remarks regarding the content of this text, a few summary pages of the ideas and concepts presented previously will be helpful. As was discussed in the opening chapter, risk communications and its sister process, crisis communications, are rapidly becoming essential skills for the toolbox of any safety, health, and environmental (SH&E) professional. Both allow for the orderly and effective transmission of information during periods of high stress to an organization. This text has focused on a number of important concepts in an attempt to provide the reader with an overall primer on the topic, allowing for further research and study as needed.

Chapter 2 provided a historical view of risk and crisis communications, noting that the term "risk communication" was first thought to be attributed to William Ruckelshaus, the first administrator of the Environmental Protection Agency (EPA) in 1970, who marshaled the organization through its first years, establishing a role in protecting the environment and assisting other community organizations in their roles. In the 1980s, the EPA's Superfund program incorporated the concept in its public participation process, and it also appeared in the Emergency Planning and Community Right to Know Provisions of Title III of the Superfund Amendments and Reauthorization Act of 1986 (Covello *et al.* 1997). Most of the roots of the theories and processes of risk

Risk and Crisis Communications: Methods and Messages, First Edition. Pamela (Ferrante) Walaski.
© 2011 John Wiley & Sons, Inc. Published 2011 by John Wiley & Sons, Inc.

communications come from the environmental arena and working with the public and other stakeholders, but in recent years the concepts have been successfully used to deal with any type of hazardous situation or disaster.

In order to establish a working knowledge for the reader of the most common terms were provided in Chapter 2. To restate examples of the two key terms, "risk communication" is, as defined by the U.S. Department of Health and Human Services (DHHS), "an interactive process of exchange of information and opinion among individuals, groups, and institutions; often involves multiple messages about the nature of risk or expressing concerns, opinions or reactions to risk messages or to legal and institutional arrangement for risk management" (U.S. DHHS 2006).

And "crisis communication," according to Fearn-Banks (2007), "is concerned with transferring of information to significant persons (publics) to help avoid or prevent a crisis (or negative occurrence), recover from a crisis, and maintain or enhance reputation."

Therefore, the critical difference is the situations in which the various communication forms take place. Risk communication is an ongoing process that helps to define a problem and solicit involvement and action before an emergency occurs, whereas crisis communications are those messages that are given to stakeholders during an emergency event that threatens them.

THEORETICAL MODELS AND FRAMEWORKS

As this text has demonstrated, much of the applicability of risk communications comes from understanding how the general public perceives risk. By understanding the perception of risk, an SH&E professional can determine how to tailor the risk message. Numerous models have been theorized and provide the SH&E professional with a framework for understanding how risk and crisis messages are perceived and were discussed in detail in Chapter 3. Vincent Covello and his colleagues at The Center for Risk Communication offer four theoretical models that help practitioners understand how information is processed, how perceptions are formed, and how risk decisions are made, all of which were discussed in depth in that chapter and applied variously throughout the rest of the text (risk perception, mental noise, negative dominance, and trust determination). By understanding these models and how they apply in various situations, SH&E professionals can better prepare their messages and coordinate their communication in high-risk situations. The key model calls upon the two important factors of trust and credibility of the communicating organization; therefore, it is critical to understand what factors can aid an organizational communicator in determining whether or not trust can be built and credibility can be achieved. In studies conducted by Covello and others at The Center for Risk Communication, three factors were determined to be key: (1) perceptions of knowledge and expertise, (2) perceptions of openness and honesty, and (3) perceptions of concern and care. The essential importance of these three factors and how they are addressed with respect to organizational trust and credibility has also been discussed at length throughout this text, particularly within the case study of the BP Deepwater Horizon oil spill in Chapter 10.

In addition to the theories noted above, the writings of Peter Sandman also provide a substantial body of work to the understanding of risk and crisis communications. His Risk = Hazard + Outrage theory is much quoted and was discussed in detail in Chapter 3 as well as in Chapter 10, where its application to two case studies (the BP oil spill, as already noted, and the H1N1 pandemic of 2010) was carefully illustrated. According to Sandman, hazards or hazardous events can be classified anywhere along the continuum from negligible to catastrophic and are assessed utilizing typical risk assessment methodologies. Outrage refers to the emotions and behaviors of the message receivers in light of their perceptions of the level of hazard presented to them. Like hazards, the level of outrage exists on a continuum from high to low. Combining those two variables provides the following types of communication situations (Sandman 2003):

1. **High hazard/low outrage (precaution advocacy).** This situation features a serious hazard, but an apathetic audience. Messages delivered in this scenario require a skilled communicator who can move an audience's outrage level to more closely match the hazard, without exaggerating it and either causing the audience to sink further into denial or apathy or lose trust in the veracity of the communicator's message.

2. **Medium hazard/medium outrage (stakeholder relations).** This audience in this situation is interested, but not so emotional that internalizing the message is difficult. It allows the message sender to discuss the situation rationally and openly and is likely to generate audience questions and rational concerns. This is the easiest communication environment and the task is to simply provide an open and honest dialogue that explains the situation and allows sufficient opportunity for audience response and questioning.

3. **Low hazard/high outrage (outrage management).** The audience is often highly outraged at what it perceives to be a serious situation and is often unreceptive to messages that counter its perceptions. Communicators must take care to give credibility to the audience's outrage and still attempt to move it to be more in line with the actual hazard. This requires sincere listening, acknowledging, and even apologizing, if that will move the audience to a more realistic view of the seriousness of the hazard.

4. **High hazard/high outrage (crisis communications).** In this scenario, the hazard is serious and the audience's outrage matches it. The communicator in this situation must tread carefully, allowing for the audience's legitimate fears, remaining "human" and empathetic but still rational, and demonstrating true leadership. The advantage for the communicator is that the outrage is not typically directed at them personally, at least until after the crisis is past.

CRAFTING RISK AND CRISIS MESSAGES

As was demonstrated throughout the text, the models above aid in the development of both risk and crisis communication messages. As important as utilizing these models is, however, is the organization's ability to accurately state the goals and objectives of the message before attempting to craft the content. This process involves establishing the goals and objectives from the top of the organizational structure, while also understanding the constraints placed upon the organization and its messages by various internal and external sources.

Organizations also need to profile the audience and understand its stake in the process or situation at hand by asking what the audience wants and needs to know as well as how the members of unique audiences understand risk. Chapter 4 provided the reader with some frameworks for both identifying audiences and for detailing their characteristics, including the concepts of four different publics as offered by Fearn-Banks (2007):

1. **Enabling publics**—publics who run organizations
2. **Functional publics**—publics who are responsible for making the organization operate
3. **Normative publics**—publics who share values with the organization and may be partners in the process of risk and crisis communications
4. **Diffused publics**—publics with a vital but indirect link to the organization

Finally, the audience profile process needs to include an understanding of the typical emotions (as Sandman would call outrage) experienced by audiences. These include anger, mistrust, fear, denial, and apathy, and each emotional state requires varying types of messages from an organization. These audience states need to be ascertained before messages are developed and in conjunction with the goals of the organizations' messages.

Once the organization knows what it hopes to achieve during the communication process and understand the audience, specific messages can be crafted. A large portion of Chapter 5 was devoted to providing an in-depth discussion of two models for message crafting: mental models, as developed by M. Granger Morgan (Morgan *et al.* 2002), and message mapping, as developed by Vincent Covello and his colleagues (Covello 2002). While the mental models approach is a much more complex process and involves the development of written materials intended for wide-scale distribution, the message mapping process can be summarized in the following seven steps:

1. **Identify potential stakeholders.** Each crisis will have a unique set of stakeholders and include external publics and well as internal ones.
2. **Identify potential stakeholder questions and concerns.** This list will also be different for each stakeholder group identified, although some common groups can be put together.
3. **Analyze questions to identify common sets of concerns.** Using a brainstorming process with an assembled group of key organizational representatives, a lengthy list of questions and concerns emerges with limited obvious patterns and should be culled down to three main concerns and associated questions that can be answered with messages.

4. **Develop key messages.** Each concern/question grouping should have three separate messages developed to address it. The messages have a very tight structure and limited content, matching typical media sound bites; therefore, the message length should be no longer than 27 words and/or be able to be read in nine seconds.

5. **Develop supporting facts.** For each of the three key messages developed in the message map, three separate supporting facts should be provided by the available research and literature. These supporting facts provide additional information for the communicator and lend credibility to the message.

6. **Test and practice messages.** Standardized procedures for message testing include asking subject-matter experts not directly involved in the communication event to validate the message content and then participate in a practice session delivering specific messages with groups that are representative of the key characteristics of the eventual intended audiences.

7. **Delivery of maps through appropriate channels.** Typical channels include various governmental agencies and the media through press conferences and releases, informational forums, internal staff meetings, community meetings, and written content that might appear on a website, brochure, or FAQ sheet.

MESSAGE DELIVERY

The actual delivery of the message in terms of the attributes of the communicator as well as the setting were addressed in Chapters 5 and 7, since even the best message will fail in the wrong setting and delivered by the wrong spokesperson. As detailed in Chapter 5, the U.S. EPA has created a list of "7 Cardinal Rules" for message delivery that are restated here due to their near universal applicability toward all types of situations and all manner of audiences (U.S. EPA 2005).

1. Accept and involve the public as a legitimate partner.
2. Plan carefully and evaluate your efforts.
3. Listen to the public's specific concerns.
4. Be honest, frank, and open.
5. Coordinate and collaborate with other credible sources.
6. Meet the needs of the media.
7. Speak clearly and with compassion.

Even the most skilled risk communicator may have difficulties in delivering the message or in finding a way in which the audience will respond as desired. As noted above, analyzing each situation carefully and understanding the audience's motivations, moods, and outrage level are essential if the message is to have a chance to succeed. It is worth summarizing some of the most common pitfalls that can derail messages that appear unable to fail (U.S. DHHS 2006).

- **Using abstractions.** Risk communicators should not assume a common understanding. Jargon, acronyms, and highly technical language should be avoided.
- **Attacking the audience.** Respond to issues, not people, and be careful to end debates by responding with clarity and factual information.
- **Sending negative nonverbals.** It goes without saying that a risk communicator who loses his/her temper is in deep trouble, but tense facial expressions and certain hand movements can also signal negativity and hostility toward the audience. Practicing in front of a mirror or another colleague can help communicators see what they may be saying nonverbally.
- **Blaming anyone.** It is never helpful to assign blame to another party in the process; it confuses audience members and forces them to take sides. Along the same lines, if an organization has some responsibility for the situation, accepting it matter-of-factly and honestly can help build trust and credibility.
- **Focusing too much on the money.** Complaining about the lack of funds to solve the problem only increases the audience's frustration since it often has no ability to change the situation anyway. It is more productive to tell the audience what benefits are being derived from the funds that are available and are being spent.
- **Providing guarantees.** Instead, the communicator should offer likelihoods and emphasize the progress that is being made or has already been made.
- **Trying to be funny.** This is usually only effective if directing the laugh at oneself. Attempting to inject humor into a serious situation unnecessarily trivializes it.
- **Going on and on and on.** Audience presentations should aim to be no longer than 15 minutes (or even shorter), while reserving plenty of time for questions. The latter can serve to effectively enhance and clarify additional message points.
- **Using negative words and phrases.** Negative messages override an audience's ability to respond and move away from high levels of emotionality. It is best to avoid them if at all possible.
- **Thinking you are "off the record."** A risk/crisis communicator never is, and nothing said to anyone, particularly the media, is confidential.
- **Promising anything.** If there is no certainty of delivery, this tactic will likely be regretted. Making strong assurances is a better move.
- **Forgetting the visuals.** Most people understand messages that are delivered in more than one format. Slides, handouts, and other visuals can enhance what is being said and helps the audience to process complicated information after the formal presentation is over.
- **Overusing statistics.** Playing the numbers game too hard is boring and is not central to the message. Statistics should be used to enhance and support comments only.
- **Forgetting to define the message goals in advance.** Nothing is worse than being unprepared in front of a large group of people who may already be distrustful. It is a sure-fire path to disaster.

- **Forgetting the role of the public.** This is a partnership. It is crucial to build trust and credibility by engaging in a dialogue.

WORKING WITH THE MEDIA AND CHOOSING A SPOKESPERSON

Successful risk and crisis communications require a platform for delivery to large-scale audiences. The media typically fill this role in print or broadcasts. As was discussed in Chapter 7, having a pre-existing positive relationship with local media can aid in the success of this process. Seeking them out in advance to offer information about an organization and being available when general stories require information and quotes can create the sense that the organization is open to the media and willing to provide information as needed.

It is often helpful if within an organization's crisis communications plan is a separate element that details media communications. Because of the potential harm done to an organization when the wrong message is delivered by the wrong person at the wrong time, this separate element helps to guide the organization's operational staff and often includes these typical elements (Hyer and Covello 2005):

- Staff roles and responsibilities, including leadership for varying types of emergencies and authorized spokespersons
- Procedures for information, verification, clearance, and approval prior to delivery to the media
- Procedures for coordinating with other partners
- Policies regarding employee contacts from media
- Media contact lists that are regularly verified
- Exercises and drills for testing media communications
- List of internal and external subject-matter experts
- Preferred communication channels (telephones hotlines, websites, radio announcements, news conferences)
- Location of repository of prepared messages
- Location of useful documents (fact sheets, FAQs, brochures)
- Task checklist for key time frames (generally at 2, 4, 8, 12, 16, 24, and 48 hours)
- Procedures for reviewing and revising the plan

Hyer also provides excellent suggestions for choosing an organizational spokesperson and offers these characteristics of an effective representative (Hyer and Covello 2005).

The designated lead spokesperson should:

- Possess excellent media skills
- Have sufficient authority or expertise to be accepted as speaking on behalf of the organization

- Possess or work to develop good professional relationships with important members of the media and other important partners and stakeholders

This person should also be:

- Perceived as authoritative and credible by stakeholders, partners, and the public
- At ease with the media
- Knowledgeable (generally and specifically) about the emergency, its dynamics, and its management
- A subject-matter expert on the event or able to delegate to subject-matter experts
- Resourceful

Finally, the lead spokesperson should be able to:

- Learn quickly
- Respond to sensitive questions within their areas of expertise in a professional and sensitive manner
- Effectively respond to hostile questions
- Stay on message yet remain flexible and able to make decisions quickly
- Offer examples, anecdotes, and stories
- Provide effective on-the-spot responses to media enquiries
- Express technical knowledge or complex information in a way that can be easily understood by reporters and by the average person
- Remain calm and composed at all times
- Express caring, listening, empathy, and compassion
- Work well under pressure or high emotional strain
- Accept constructive feedback
- Share the spotlight
- Call on the expertise of others
- Give thanks to others and distribute praise
- Take responsibility for things that go wrong
- Present the appropriate tone for the audience
- Defer, delegate, and redirect questions to others as needed

Not all management representatives are able to fill this role, and even those who can may need support in various situations such as when the information is highly technical or when an incident drags out over an extended period of time and multiple spokespersons are needed to take on the task.

Finally, some general rules for spokespeople who will be interviewed by the media are offered as compiled from various sources (Donovan and Covello 1989; Fearn-Banks 2007; Hyer and Covello 2005; Lundgren and McMakin 2004; U.S. DHHS 2006; Hurns and Tapp 2010):

- Be patient, open, and honest
- Listen carefully to the question and think before you speak
- Don't feel obligated to answer every question
- If you begin to feel uncomfortable with the directions of the interview, return to your key messages
- Avoid opinions or speculation
- Use common language
- Don't speak disparagingly of others
- Be prepared to support your facts
- Know what controversial information has already been published
- Don't take anything personally
- Don't embarrass the reporter or argue with them
- You are never off the record
- The reporter gets to pick the topic and questions
- Know what topics are off limits for you
- Never say "no comment"
- If you make a mistake, ask for the chance to clarify
- Check your quotes

DEVELOPING A RISK/CRISIS COMMUNICATIONS PLAN

As is noted in Chapter 8, the key rationale for developing a risk/crisis communications plan (hereafter referred to as the "Plan") is to anticipate risks, assess and evaluate them, and then determine, in advance, how the organization will respond to these risks, specifically with regard to communications. Risk assessment techniques were summarized in Chapter 8, and included matrices, checklists, and "what-if" analyses. (These topics were covered in a limited fashion, and the reader was referred to other more specific documents for further study of this area.)

As noted previously, the Plan is also designed to use risk communications to develop the trust and credibility with the targeted audience in advance of any crisis eruption and dictates the processes that will be followed when crisis communications are required. A model planning tool developed by the Texas Department of State Health Services is presented below. As with all organizational planning processes, all levels of management and operations must be involved in the process and participate on the working group. Sufficient authority must be provided to the planning and implementations processes and include resources such as money and time. The planning process will ideally be team-focused and may also include external resources and groups.

Key elements of the Plan include (Reynolds 2002):

- Signed endorsement from top levels of management
- Designated line and staff responsibilities for all levels of the organization

- Internal information verification and clearances/approval procedures
- Agreements on information release to authorities
- Designated spokesperson for audiences as well as validation for third-party responders
- Regional and local media contact list (to include after-hours news desks)
- Procedures for coordinating internal response teams
- After-hours contact list and contact information for internal and external sources
- Agreements and procedures to join a joint information or emergency response center
- Procedures to secure needed resources (space, equipment, people, finances) to operate during an emergency 24 hours a day/7 days a week if needed
- Identified vehicles of information dissemination to audiences

Drawing directly from several planning tool references, below is a working Plan outline:

    I. Authority
   II. Purpose
  III. Scope
  IV. Situations and Assumptions
   V. Concept and Operations
  VI. Organization and Assignment of Responsibilities
 VII. Plan Development and Maintenance
      Appendices

SPECIAL RISK AND CRISIS COMMUNICATIONS SITUATIONS

Chapter 9 devoted itself to a more detailed study of the process for dealing with special risk or crisis communications situations that are more likely to occur for a majority of organizations and because they provide a means of utilizing the previous information in the text and applying it to a specific situation. Those chosen for were "worst-case scenarios," fatalities, and rumors.

The discussion of worst-case scenarios focused more on technique and provided recommendations offered by Sandman called "moving the seesaw" (Sandman 2001). The worst thing that can happen regarding a particular risk event is most often a low-probability event but also one that can cause opposite responses in an audience—apathy and extreme outrage—because of its high level of severity. "Moving the seesaw" refers to the need to identify an audience's position and take the opposite one, causing the audience to move toward the other position.

Fatalities are situations in which the outrage level is often extremely high, and for good reason. Communications, particularly with the workforce, need to address the outrage level and acknowledge it. Specific concepts about message crafting and delivery are summarized below (Clausen 2009; Sandman 2004):

- Any attempt to downplay the situation will be met with justifiable anger and mistrust. Attempts to over-reassure the workforce that the situation should not be cause for sadness and anger, particularly in the early stages, are at the best ineffective and at worst, callous.
- The communicator will be relying heavily on previously built levels of trust and credibility as the initial messages are delivered. It goes without saying that a workforce with a fundamentally sound level of trust in its leadership will be more receptive to messages during this type of crisis than one in which the relationship between management and the workforce is adversarial.
- Once the message has been delivered that details the basic information about what has happened, members of the workforce will often benefit from being given some task that they can perform. Action binds anxiety, and it can be assumed that the workforce will be suffering high levels of anxiety in the early minutes, hours, and sometimes days following a fatality.
- Panic in an audience is not a typical reaction, even in the face of extreme danger. When given accurate information and actions to be taken to protect themselves and others, most audience members will do what is in their best interests and, for the most part, what is in the best interests of the group.
- Communicators should be instructed to be willing to share their own pain during communication events that discuss the fatality. Acknowledging the personal impact the situation has had on the individual communicator does not make them less professional or ethical; it makes them more human.
- It is critical that communications occur early and often, even in the absence of much information about what happened. The vacuum created by lack of timely information can be detrimental as workers will fill that vacuum with their own ideas and information.
- When a fatality occurs, the ability of the spokesperson to express compassion and concern is heightened. While many communication situations require someone who can translate complicated concepts into understandable messages, the demeanor of the spokesperson in a crisis like a fatality is more important, particularly in the early stages.
- In recent years, more has been learned about the circumstances under which post-traumatic stress disorder (PTSD) may occur. The possibility of seeing this type of reaction exists along the continuum from mild to severe among members of a workforce. As soon as the communication events begin regarding the incident, concurrent actions should also begin to provide for the psychological mental health of the workforce and associated audience members.

With regard to rumors, they often precede a crisis for an organization and, in some cases, precipitate one, particularly if the organization is unprepared or its reaction to them is flawed. A framework for classifying the types of rumors was provided by Fearn-Banks (2007):

1. **Intentional.** This rumor is intentionally started by an organization with a purpose and can be the type of rumor deliberately started to achieve sales

growth such as is often done by those in the financial industry to increase the prices of stocks or mutual funds.

2. **Premature fact.** In the beginning, these rumors do not appear to have a basis in fact, but eventually do turn out to be true.

3. **Malicious.** Like intentional rumors, these are deliberately started, but the purpose is to harm, and unlike intentional rumors, these are started by a competing organization.

4. **Outrageous.** The content of this type of rumor is essentially unbelievable; however, the response from those who hear this rumor ends up being that it must be true because it couldn't be made up.

5. **Nearly true.** Portions of this type of rumor are true, but not all of it, unlike the premature-fact rumors where the entire content appears untrue.

6. **Birthday.** These rumors continue to emerge over and over again, just like a birthday.

Determining the best strategy for dealing with a rumor is complicated by several unknowns that are typical; the organization doesn't know how the rumor started and how far it has spread. Further complicating strategies is the decision about whether or not to respond to the rumor, which can increase the likelihood of it spreading, or to do nothing, expecting that it will die in time.

General proactive strategies for dealing with rumors, also offered by Fearn-Banks, include the following:

- Organizations can be the last to know about a rumor or find out about it after it has been circulating for quite some time, limiting the strategies available to deal with it. Mechanisms for audiences to provide information can help with early detection.

- Running periodic focus groups that query a variety of audiences about their perspective of an organization can turn up information generally known and accepted by the audience but not the organization.

- Management of other organizational representatives should be trained in rumor detection and control. Front-line supervisors are often the recipients of valuable information and hear rumors either directly from the workforce or in bits and pieces as they perform their job duties.

- Organizations should make it a practice to pay close attention when a similar type of organization is the subject of a rumor and observe how the organization manages it, as well as the outcome of the strategies employed. These types of prodromes provide valuable information that can be used in future efforts.

- The lines of communication with key audiences should be well developed, allowing for information to flow out when possible problems erupt or are on the horizon.

- An organization should ensure that it has a good deal of trust and credibility with its key audiences.

Finally, strategies for dealing with rumors that have already begun to circulate include the following (Fearn-Banks 2007; U.S. DHHS 2006):

- Time is a critical factor in rumor response; a previously developed plan with a set of organizational goals and objectives, spokespersons, and critical message templates will jump-start any response.
- The organization will need to be able to quickly ascertain the extent of the spread of a rumor, with the understanding that literally each hour that goes by increases the chances that the spread of the rumor will grow. This is important to know because the extent of the spread will help determine the proper response. Not every rumor demands a full frontal attack.
- Effective rumor control occurs when the organization accurately predicts how a rumor might evolve. Knowing your audiences through focus groups, open lines of two-way communication paths, and a risk communication strategy that is regularly implemented and evaluated will provide information.
- An organization needs to ensure that messages delivered to counteract or dispute rumors are clear and focus only on factual information.
- Messages delivered regarding rumors should offer information that is contradictory to the rumor's content without ever mentioning the rumor.
- There are some situations in which doing nothing after becoming aware of the existence of a rumor is an effective strategy. This would be true if denying the rumor will only bring more attention to it.
- The opposite strategy of loudly and very publicly denying the rumor works best when the organization has proof that is easy for the average person to verify that the rumor is not true. It is also helpful when taking this route to ensure that a positive relationship with the local media is already in place, as they will not only provide a method for spreading the organization's response but can also be expected to verify the offered facts in their reports.
- In order to distance the organization from the rumor, a credible spokesperson from outside the organization, a celebrity or well-known public figure such as a university professor, can be called upon to discredit the rumor in some public manner (e.g., news conferences, media interviews).
- Advertisements or notices in publications with widespread circulations or distinct circulations within the audience most likely to be spreading the rumor are effective ways to reach large number of people.

CASE STUDIES

As a means to bring the major concepts and theories of the text together, two in-depth case studies were offered at the end of the text. These case studies, the H1N1 pandemic of 2009–2010 and the BP Deepwater Horizon oil spill of 2010, offer the opportunity to analyze messages delivered by various organizations involved in the crisis. The messages presented in the case studies were culled from various media sources including print and broadcast journalism as well as press releases and transcripts of public meetings. The successes and failures of the messages were highlighted along several key themes such as ability to successfully move the outrage level of the audience, the success and failures of key spokespersons, and the success in the message delivery with regard to audience behavior and attitude change.

While few organizations will encounter a crisis with the magnitude and length of coverage experienced by BP, and the H1N1 pandemic was a worldwide event with hundreds of different organizations participating in the messaging, both case studies provide excellent opportunities to analyze, after the fact, the effects of the messages and how the crises unfolded in part as a result of those messages.

WHAT IT ALL MEANS FOR YOU AND YOUR ORGANIZATION

As SH&E professionals, our roles within organizations as internal consultants have evolved dramatically over the past 25 years. We have had to adjust our skill set and knowledge base to encompass more and more areas, typically environmental regulations and more recently sustainability, and are called upon more often to become quasi-experts. Risk and crisis communications are another example of these new areas.

While some SH&E professionals may rarely, if ever, be involved in risk and crisis communications, many more will, and that number is expected to grow as the SH&E professional task list grows, and the need to continue to add value to organizations grows. And as the information delivery system in our culture continues to expand and involve more and more frequent messages through multiple channels, more and more and more organizations will be called upon to deliver risk and crisis messages and will need professional expertise within the structure of their management and technical staff. The tasks SH&E professionals may be called upon to do include assisting with audience profiling, message crafting, and even occasionally delivery, particularly as the front-line communicators to one significant organizational audience—the workforce. In fact, SH&E professionals might do well to look closely at the concepts of risk communications and utilize them to help develop curricula for all manner of safety training, which is, in a very real sense, risk communications. The concept of audience profiling might help solve dilemmas posed by indifferent attendees, and message mapping might better address the true concerns and questions that would make the time in training more memorable for all involved.

Texts such as the one just presented attempt to provide SH&E professionals with fundamental information that can be used to begin to develop the rudimentary skills and expertise needed to provide assistance within their organizations. The basic concepts covered herein, while nearly universal in their applicability, provide only a fraction of the overall process of risk and crisis communications. More in-depth study will help develop a broader range of skills and abilities if needed, and the reader is referred to the many sources quoted throughout this text as a means to begin.

REFERENCES

Claussen, L. 2009. "After the Incident: How to Deliver the Message to Employees and Family Members About Workplace Victims." *Safety+Health* 180(5):48–51.

Covello, V. 2002. "Message Mapping, Risk and Crisis Communication." Invited paper presented at World Health Conference on Bio-Terrorism and Risk Communication, Geneva, Switzerland, October 1.

Covello, V., R. Peters, and D. McCallum. 1997. "The Determinants of Trust and Credibility in Environmental Risk Communication." *Risk Analysis* 17(1):43–54.

Donovan, E. and V. Covello. 1989. *Risk Communication Student Manual.* Washington, DC: Chemical Manufacturers Association.

Fearn-Banks, K. 2007. *Crisis Communications: A Casebook Approach,* 3rd ed. Mahwah, New Jersey: Lawrence Erlbaum Associates.

Hurns, D. and L. Tapp. 2010. "Working With the Media: Telling Your Story Effectively." *Professional Safety* 55(1):52–54.

Hyer, R. and V. Covello. 2005. *Effective Media Communication During Public Health Emergencies.* Geneva, Switzerland: World Health Organization.

Lundgren, R.E. and A.H. McMakin. 2004. *Risk Communication: A Handbook for Communicating Environmental, Safety, and Health Risks,* 3rd ed. Columbus, OH: Battelle Press.

Morgan, M.G., B. Fishhoff, A. Bostrom, and C.J. Atman. 2002. *Risk Communication: A Mental Models Approach.* New York: Cambridge University Press.

Reynolds, B. 2002. "Crisis and Emergency Risk Communication." U.S. Centers for Disease Control and Prevention, Atlanta, GA.

Sandman, P. 2001. "Advice for President Bartlet: Riding the Seesaw." Posted online at www.ptersandman.com/col/westwing.htm on July 14, 2001. Accessed on October 23, 2010.

Sandman, P. 2003. "Four Kinds of Risk Communication." Posted online at www.petersandman.com. Accessed on November 1, 2010.

Sandman, P. 2004. "Worst Case Scenarios." Posted online at http://www.petersandman.com/col/birdflu.htm on August 28, 2004. Accessed on August 6, 2010.

U.S. Department of Health and Human Services. 2006. "Communicating in a Crisis: Risk Communication Guidelines for Public Officials." Washington, D.C.

U.S. Environmental Protection Agency. 2005. "Superfund Community Involvement Handbook." EPA 540-K-05-003.

INDEX

Risk and Crisis Communications: Methods and Messages, First Edition. Pamela (Ferrante) Walaski.
© 2011 John Wiley & Sons, Inc. Published 2011 by John Wiley & Sons, Inc.

Printed and bound by CPI Group (UK) Ltd, Croydon, CR0 4YY

27/10/2024

14580260-0001